Thylacine: The Tragedy of the Tasmanian Tiger

Thylacine:

The Tragedy of the Tasmanian Tiger

Eric R. Guiler

Melbourne
Oxford University Press
Oxford Auckland New York

OXFORD UNIVERSITY PRESS

Oxford London New York Toronto
Delhi Bombay Calcutta Madras Karachi
Kuala Lumpur Singapore Hong Kong Tokyo
Nairobi Dar es Salaam Cape Town
Melbourne Auckland
and associates in
Beirut Berlin Ibadan Mexico City Nicosia

National Library of Australia
Cataloguing-in-Publication data:
Guiler, E.R. (Eric Rowland), 1922–
    Thylacine, the tragedy of the Tasmanian Tiger.

    Bibliography.
    Includes index.
    ISBN 0 19 554603 2.

    1. Thylacinus cynocephalus. I. Title.

599.2

Edited by Jill Taylor
Designed by Guy Mirabella
Typeset by Asco Trade Typesetting Limited Hong Kong
Printed in Hong Kong
Published by Oxford University Press, 7 Bowen Crescent, Melbourne
OXFORD is a trademark of Oxford University Press

Contents

Preface

The thylacine, or Tasmanian tiger as it was called, has attracted a great deal of attention not only in recent years but also during the time when it was still relatively abundant. Its carnivorous habits are alien to the major group of animals to which it belongs, and very different from the gentle nature associated with the kangaroo and the koala. It is this difference in its habits together with its appearance that has set it apart from other marsupials in the public mind. The image of a lonely carnivore stalking the bush in search of food and a mate, whilst every hand is turned against it, has led to a fascination which has been increased over the years by the haunting question of whether or not the animal still exists.

Concern for the animal has not been confined to Tasmania or even to Australia but has extended overseas, and news of thylacine searches receives worldwide cover. Scientists, naturalists, and the whole spectrum of that amorphous mass called 'the public' have all expressed their interest in and concern for the thylacine. Needless to say the media have pursued the topic from time to time with varying degrees of success and responsibility, and articles have been written sometimes with more enthusiasm than accuracy. All of these have served to stimulate interest in this elusive animal.

My own involvement with the thylacine started about

1958 when I was chairman of the Animals and Birds Protection Board (referred to hereafter as the Fauna Board) and it became apparent that little was being done to investigate the alleged sightings which were reported from time to time. The Fauna Board, which was the government statutory authority controlling faunal matters in Tasmania, agreed to investigate all new sightings and reports. In order to determine something of the background to this project I carried out a survey of what was then known of thylacines. After this I became involved in official government expeditions organized by the Fauna Board to try to catch a thylacine, and later ran World Wildlife Fund investigations intò whether the species was still in existence. My most recent involvement has been a World Wildlife Fund (Australia) photographic quest to try to obtain a photograph of a thylacine (see Chapter 8).

In the early 1960s, as part of the historical survey, I interviewed most of the few surviving trappers who had actually caught thylacines and gained from them first-hand information about their experiences. The last year in which thylacines were relatively common was 1908, and active thylacine trapping had practically ceased by 1912. Most of the 'tiger tales' of today are at best second-hand, being based upon recollections of fathers' or grandfathers' reminiscences. There are one or two old men still living who caught one or even a few thylacines when they had become scarce after 1908, but unlike the trappers in the bounty days they had little real experience of the animal and its ways. All the old trappers who caught perhaps a dozen or more thylacines are dead now, taking their knowledge with them but leaving behind tales of the animal and its habits, some of which are recorded in this book. I talked to many old-timers, men like George Wainwright who was the last 'tiger man' on Woolnorth, Joe Willett of Ross, and H. and T. Pearce of Derwent Bridge, who between them caught dozens of 'them useless things'.

The scientific literature on thylacines is widely scattered and most of it has been brought together in the chapters on the animal and its biology. These represent the start and finish of our scientific knowledge of the animal. I have read much of the popular literature and it is even more scattered. Some is anecdotal or inaccurate, or both, or repetitive and I rejected it, but relevant passages are quoted where appropriate. In addition I have received letters relating to thylacines, and where these are quoted and referenced the relevant letter has been deposited in the State Archives of the State Library of Tasmania. Interviews and conversations are referred to in the text as (pers. comm.).

I have decided to present all of my notes and experiences in the form of a book since much of the material does not conform to the strict and rather restrictive demands of scientific journals. Also, these journals are not widely available and I believe that there are many people in the community who would like to know more about this elusive animal.

I have tried to make this a comprehensive study, but so little is known of the thylacine other than some historical and anatomical information that I have had to draw certain conclusions from the material available, and some of this has not been published before. In the course of the background reading for this book I have come across much of interest relating to the early times and country life of Tasmania, and sometimes I have included a little of this if it seemed relevant.

No such work as this could be a single-handed effort and I want to thank all those who have helped me in various ways as well as those persons who have taken the trouble to write to me.

The frequently arduous early field work was largely carried out by the late Inspector G.J. Hanlon BEM, Tasmania Police, senior wildlife officers R. Hooper and K. Harmon, wildlife officer K. Norton, and Mr R. Martin of the Fauna Board. In more recent times this work has been done by Mr N. Mooney and Mr G. Melville of the National Parks and Wildlife Service.

Most of our field equipment was maintained by Mr R. Wheeldon who also built it to the designs of Mr C. Williams, both of the Zoology Department of the University of Tasmania. My friend Mr J. Reynolds of Zeehan serviced our gear for a year or so. The construction of the equipment to capture, photograph, or otherwise entice thylacines was constructed in the workshops of the Zoology Department. These persons have borne the brunt of the demands of the project and I am very grateful to them as without their efforts there would have been no project.

The archival research has meant many hours in the State Archives since 1960 and I am grateful to the staff who have searched or carried many tomes for me. The first archival research was carried out at both Woolnorth and Burnie where I had access to the Van Diemen's Land Company records and I thank Mr P. Busby for permission to search this magnificent material and also for his hospitality whilst at Burnie and Woolnorth.

I am particularly grateful to Mr G. Roberts for permission to peruse and quote from the diaries and cash books of the Beaumaris Zoo which was owned by his grandmother, Mrs M. Roberts, and to the Hobart City Council for the same permission in relation to the later records of the Zoo when it was run by the City Council.

All of this work would not have been possible without financial support and I am very pleased to acknowledge the backing that I have received from the Tasmanian Government and from the Fauna Board as well as from its successor the National Parks and Wildlife Service. The World Wildlife Fund provided support for two searches and World Wildlife Fund (Australia) funded the last major quest. BP Australia made me a grant for the Mark II cameras. I am very grateful to them all.

Inevitably, since all of this book was written at home, there has been more than some disruption to home life as table space has been at a premium and I sincerely thank my wife for putting up with it all as well as for reading the drafts to try to make them make sense.

I have enjoyed the company of my field companions and my chats with landowners, old-timers, and others who have given me freely of their experience, friendship, and hospitality. I have enjoyed the bush, the camaraderie of the campfire, the debates, yarns, the hopes and even, in an odd sort of way, the disappointments, frustrations, and the failures. I have seen many remote and beautiful parts of Tasmania which I certainly would not have visited otherwise. All this adds up to many happy, often tiring, frequently wet hours servicing gear or searching in all weathers for this so, so elusive animal.

I am still searching.   Eric Guiler
                                                         Hobart 1984

Didelphis fusco-flavescens supra postice negro-fasciata, cauda compressa subtus lateribusque nuda

Harris, 1808.

1
Introduction

Tasmania, or Van Diemen's Land as it was then called, was settled by Europeans in 1803. The Tasmanian Aborigines had been there for about 50 000 years before that, but they had never been very numerous. The first years of the colony were characterized by essentially local settlement around the two loci of Hobart and the Tamar Valley, but the colony expanded steadily and by 1820 Hobart was the second largest town in Australia (Blainey 1966). The earliest major industries in the colony were directed seawards—seeking whales and seals—but the land, particularly the fertile plains and coastal regions, was rapidly explored and surveyed and made available to settlers. This activity was mainly agricultural in emphasis and brought the colonists into contact with the local fauna.

Among the first settlers at Hobart was the official padre to the colony, Rev. R. Knopwood. His diary is an account of the trials, events, and successes of the infant colony. He wrote:

18 June, 1805. Am engaged all the morn. upon business examining the 5 prisoners that went into the bush. They informed me that on 2 May when they were in the wood they see a large tyger that the dog they had with them went nearly up to it and when the tyger see the men which were about 100 yards away from it, it went away. I make no doubt but here are many wild animals which we have not yet seen. (Knopwood 1805)

Was this the first sighting by a European of a thylacine? Surely there is something tragic about those five unknown convicts who absconded into the alien bush and were the first to see a thylacine, only to return to captivity and an almost certain flogging for their adventures.

Knopwood's report was preceded by that of Paterson (1805), whose dogs killed an animal 'of carnivorous and voracious tribe' on 30 March 1805 near Yorktown on the Tamar River. Beyond some further generalities Paterson neglected to describe the animal in detail and his description is so vague that it is legitimate to wonder whether what he saw was a Tasmanian devil or a thylacine. However, no matter which report is accorded priority, the year 1805 at the latest saw the thylacine meet with its ultimate tormentor.

The name 'tiger' has remained with the animal ever since, although in more recent times it has been called the 'Tasmanian tiger'. It is frequently called the 'thylacine' although this name was used more by scientists and urban dwellers than by rural people, who gave it a wide variety of names which are listed later (Chapter 2). To avoid any possible confusion with the feline tiger I shall use *thylacine* as the common name except when quoting from the literature or letters.

Thylacines did not occur in large numbers in Tasmania during the early colonial era and consequently it always was a matter for record or comment when one was seen or trapped. We have a widespread but by no means comprehensive literature as well as folklore upon which to draw for sources. In recent times when the thylacine has almost disappeared, interest and speculation about the animal has increased rather than diminished. There is no other species in the Australian fauna today which has aroused so many stories and even created its own legends in the short space of years since white settlement. The 'bunyip' has its followers and adherents in Australia, but it is a truly legendary animal, not being described in scientific literature, so it is in a different category to the thylacine.

It has always amazed me how people will fasten upon something unknown and develop a whole aura of 'authentic facts' around it to explain it in terms which we can all understand. The unknown attracts attention but it remains a mystery only so long as it is unidentified and the experts are put under pressure to rationalize the problem. Once the object is identified it loses its fascination. Some years ago I was a member of an expedition organized to investigate a stranded 'something' on the west coast of Tasmania. The object had been described as a monster larger than any animal known to us and covered with fur. This notion of a large furry monster appealed to the imagination of the public and, more importantly, that of the press which gave it front page coverage almost throughout the world. Fanciful theories were advanced as to what it could be, reaching as far back in time to an ichthyosaur or plesiosaur which had been frozen into a Jurassic ice sheet and which had melted out just now for our enlightenment. How easy it is to build up fables around the unknown. In reality the monster turned out to be a piece of whale blubber about 2½ metres long.

One of the features of the monster episode was the impression that there were thousands of people all around the world desperately hoping for such an animal to appear. Perhaps this is an atavistic hangover from our Pleistocene ancestors and the days when enormous and dangerous creatures were around every corner.

The thylacine almost fits into the same category as the monster—it is rare, if not large it is allegedly fierce, it is a predatory beast, and there are many tales about it which stir the imagination. Some of these stories are true, some might be, and others are clearly fictitious. However, the thylacine is not in the monster category as it is not unknown, it has been seen in living memory, and whole specimens are to be seen in museums.

However, whether extinct or not, the thylacine will remain for ever on record in Tasmania, since it forms one of the supports of the state coat-of-arms as well as of the coat-of-arms of the recently created city of Glenorchy. The city of Launceston has gone one better and has two thylacines as supports for its armorial bearings. It is particularly appropriate that the Royal Society of Tasmania included a thylacine in its coat-of-arms as this society was very active in investigating these animals during the decade 1840–1850.

The use of an extinct species in armorial bearings can be seen in other countries such as Mauritius, where the dodo forms one of the supports. Perhaps the nagging guilt for the extermination of a species is appeased by immortalizing the creature in this way. We can all hope that the use of the thylacine in this fashion is not a forecast of its extinction.

The name of the thylacine is further perpetuated by the Tiger Hills and Tiger Creeks scattered throughout the state, S.J. Smith (1980) listing no fewer than twenty-nine such places. Indeed, Corinna, a small settlement of holiday shacks on the west coast, is named for the thylacine using the east coast Aboriginal dialect name for the purpose.

The objective of this book is to gather together from the various sources all the information we have about the thylacine, to consider this in relation to our knowledge of other carnivorous species, and hence build up a picture of how the thylacine lived. There are many gaps in our knowledge and we shall not know a great deal about the life of the thylacine until we can study a wild population of the animals.

The geographical scene

Before examining the story of the thylacine it is necessary to set out briefly the geographical scene for the chase. Only very general features are given here and more detail can be obtained from sources such as J.L. Davies (1965).

Tasmania is the smallest state of Australia and comprises most of the islands off south-eastern mainland Australia. The small islands of Bass Strait need concern us no further as they have not supported thylacines during historical times or, as far as we know, earlier.

The island state lies in the Roaring Forties from 40° S to 43°50' S latitude and is some 68 300 square kilometres in area, being about the same size as West Virginia and slightly smaller than Ireland. It is a hilly country with open grass-lands found only in the midlands, along parts of the coast, and in the river valleys. The central plateau (over 700 m above sea level) occupies most of the middle of Tasmania, the western parts forming the mountain ranges which extend to the south coast. These mountains, as mountains go, are not high, few peaks exceeding 1500 m, but they are noted for their rugged nature, with gullies and river valleys which are very steep and covered in thick rainforest, impassable in places.

Thylacine country, Rossarden, 1958. Note the absence of well defined trails in this open bush.

The topography of the mountains make progression difficult for humans but the central plateau and the woodlands which cover most of the rest of the state allow relatively easy access to any person prepared to walk. The central plateau largely consists of open areas of sedgeland with stands of forest on the higher ridges. This type of country is rich in marsupials, both herbivorous and carnivorous.

The coastal plains supply food and shelter for large numbers of mammals. These plains are not extensive, usually being only a few hundred metres wide.

The woodlands covering much of the country offer good shelter and food and are utilized by most of the native species. However, this habitat has been greatly affected by land development projects, particularly in relation to woodchipping. Whilst this may not be totally disastrous, it can be very serious in places where the forest is not permitted to regenerate.

The mountains, forests, and sedgelands offer effective barriers to the spread of some species so that we find that the brush possum, *Trichosurus vulpecula*, although widespread, is absent from some parts of the southwest (Hocking & Guiler

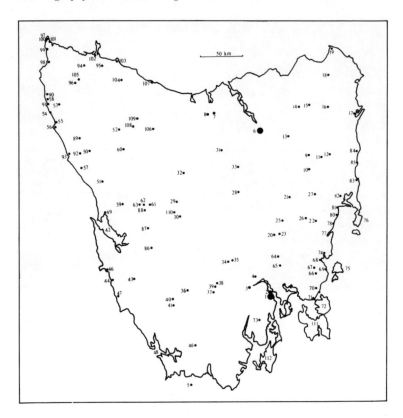

1983), and that the introduced European rabbit has been very effectively restricted in its spread by these forest-sedgelands. The forests would have been very much less effective barriers if the marsupial fauna had evolved an elephant-sized form in modern times which could barge its way through the vegetation and make tracks for other animals as well as light-breaks to enable grasses to grow and so provide food for herbivores. One of the features of these rainforests is the poverty of the food available and this is shown in the small number of larger mammals to be found there (Hocking & Guiler 1983).

The climate over most of the state is moderate without extremes of temperature and an annual rainfall of about 600–700 mm; but the central plateau and the mountains have cold weather, snow being frequent in winter, and heavy rain can

Fig.1.1 Outline map of Tasmania showing localities mentioned in the text.

Adamsfield 36; Arthur's Lake 28; Arthur River 90; Avoca 10; Balfour 53; Bicheno 83; Black Bluff 106; Blessington 13; Blue Tiers 16; Britton's Swamp 96; Broadmarsh 4; Bruni Is. 112; Bub's Hill 63; Buckland 67; Burnie 107; Campania 65; Cape Grim 100; Cape Portland 19; Cardigan River 62; Chain of Lagoons 84; Cockle Bay 69; Colebrook 64; Collingwood River 61; Corinna 92; Cranbrook 82; Cuckoo 14; Dawson River 58; Deloraine 7; Derwent Bridge 110; De Witt Is. 5; Donaldson River 89; Dunalley 71; Fingal 12; Fitzgerald 38; Florentine Valley 37; Forestier Peninsula 72; Freycinet Peninsula 76; Frenchman's Cap 87; George's Bay 17; Glenorchy 2; Golden Valley 31; Granville 57; Green Hills 102; Green's Creek 54; Hamilton 35; Hampshire Hills 108; Harcus River 101; Hobart 1; Interlaken 23; Irishtown 94; Jane River 86; Kelvedon 80; Lake Gordon 40; Lake King William 30; Lake Leake 27; Lake Pedder 41; Lake St Clair 29; Launceston 6; Lemont 24; Lisdillon 78; Little Swanport 77; Macquarie Harbor 42; Mainwaring River 47; Malahide 11; Maria Is. 75; Mawbanna 95; Mayfield 79; Mersey Lea 8; Milabena 104; Middlesex Plains 108; Moore's Valley 43; Mount Cameron West and Mount McGaw 98; Mount Victoria 15; New Norfolk 3; Oatlands 20; Orford 68; Ouse 34; Pedder River 55; Pelverata 73; Pieman River 93; Port Davey 48; Pony Bottom 66; Que River 60; Queenstown 59; Ragged Tier 70; Raglan Range 88; Ringarooma 15; Rocky Cape 103; Ross 21; Rossarden 9; St Peter's Pass 25; Sandy Cape 56; Seymour 85; South Mount Cameron 18; Spiro Range 45; Spero River 46; Strahan 49; Studland Bay 99; Surrey Hills 108; Swansea 81; Tanina 4; Tasman Peninsula 111; Temma 91; Tooms Lake 22; Triabunna 74; Trowutta 105; The Island 26; Tyenna 39; Valentine's Peak 109; Walls of Jerusalem 32; Wanderer River 44; Waratah 52; Welcome River 101; Western Tiers 33; Whyte River 50; Woolnorth 97; Zeehan 51.

fall at any time of the year. The annual precipitation can reach 3000 mm in some places.

The state's population concentrations are Hobart, Launceston, and along the northern coast. The total population is about 410 000, a density of about five to six persons per square kilometre, but this is not a very meaningful statistic since most of the south west is unpopulated and the population of the central plateau is low for most of the year, being significant only in restricted areas during the hunting or fishing seasons or in the summer. Many parts of the highlands now have fewer permanent residents than fifty years ago.

Tasmania, even today, offers a habitat which has most, if not all, of the living requirements for thylacines. It has abundant food over much of its woodland, central plateau, and coastal plain habitats. It offers seclusion over large areas which in turn give security and quiet for breeding. The absence of large human populations means that a thylacine could spend much of its life undisturbed by interference from people. Many thousands of acres of land formerly used by thylacines have been turned over to pasture but there still remains sufficient habitat for a thylacine to live undisturbed and without having to shift to a new area on account of depletion of habitat.

The localities mentioned in the text are shown on Fig. 1.1; for further details the reader is referred to the Tasmaps Series.

2

History of the thylacine

The order Dasyuroidea, of which the family Thylacinidae is a member, is found only in the Australasian biogeographic region but its fossil history is not very well known. Representatives of the Dasyuroidea have been found as far back as the Late Oligocene, but their earlier fossil history is conjectural. However, this not directly relevant to the fossil history of the thylacines since members of the Thylacinidae are not known earlier than the Miocene, and their actual relationships and origins are obscure.

Although there are many gaps in the fossil record it is evident that the species once was widespread on continental Australia, extending north to the island of New Guinea and to Tasmania at the other extreme. Fossils are well known from Recent fossil deposits in Victoria (Gill 1953, 1964), South Australia (Archer 1971), Western Australia (Glauert 1914), and north-west Australia (Kendrick & Porter 1973). The species was reported from New Guinea by van Deusen in 1963.

Considerable fossil or subfossil evidence of thylacines has been found in caves on the Nullarbor Plains of Western Australia where a number of fossil thylacines have been located

(Howlett 1960; Cook 1963; Lowry & Lowry 1967; Partridge 1967) in addition to one remarkable mummified specimen (Lowry & Merrilees 1969). The dating of these remains shows that about 7000 years BP (Archer 1974) there was a widespread thylacine and Tasmanian devil fauna on continental Australia as well as on the then separated island of Tasmania (MacIntosh 1975), and fossil thylacines have been dated to as recent a time as 3280 ± 90 BP (Partridge 1967). This fauna continued to exist until sometime after 3000 BP when both species began to disappear over continental Australia, the most recent dating being a thylacine found with dog remains some 2200 ± 96 years BP. The mummified carcases from Western Australia have been dated as 3280 ± 90 BP, with older specimens from 5440 to 4660 years BP (Merrilees 1970).

However, assuming that thylacines have vanished from all of their former range in continental Australia, we must seek some reason for their disappearance. The climate about 6000 years ago was moister than at present and started to become drier about 5000 years ago (Bowler *et al.* 1976), but it has not changed much in the last 3000 years. The moist period would have suited thylacines better than the present arid conditions, but thylacines are known to have survived well into this arid period in continental Australia. The climatic alterations, such as they were, do not lead us to conclude that they were responsible for the extinction of both the thylacine and the Tasmanian devil in continental Australia and we must look to other factors such as interspecific competition. Indeed, Ingram's paper (1969) suggests that the vegetation as long ago as 5000 years BP was very similar to that found on the Nullarbor Plains nowadays, consisting of mallee, *Eucalyptus gracilis*, *Melaleuca*, and *Acacia*. This would have been the type of habitat around the caves in the Nullarbor where the remains were found. Throughout this time thylacines were living in Tasmania.

There are three species which offered possible competition to the thylacine. Two of these are canids, the dog and the dingo, and the third is human.

The dog is believed to have come to Australia with the Aborigines perhaps about 20 000 years ago, although Mulvaney (1969) suggests a more recent date of about 8600 ± 300 BP. Dogs survive well in the wild and the moister climate of the time would have suited them. Dogs lived sympatrically with thylacines and, assuming that the dog was of large size, would compete actively with thylacines. Dogs have a very effective pack-hunting predatory technique. They also have a wide food spectrum, eating carrion and even vegetation if necessary, and would have been capable of killing thylacines (Calaby 1971). Dogs would compete with thylacines for food but have a distinct ecological advantage as thylacines do not eat carrion or vegetation and are not known to use any form of pack hunting.

Wood Jones (1921) suggested that the dingo arrived in Australia by sea, and MacIntosh (1975) suggested that this event occurred sometime after 8000 years ago, although dingoes do not appear in the archaeological context until 3000 BP. The arrival must have occurred after the separation of Tasmania and the other islands from Australia as dingoes have never been found in Tasmania. The origin of the dingo is uncertain but it is believed not to have evolved from the dog. Dingoes would have offered strong competition to thylacines in the same way as outlined in the case of the dog.

The early history of Aborigines on the continent of Australia suggests that the first arrivals occurred about 20 000 years ago, although the recent discoveries in caves in southwest Tasmania have shown that humans lived there some 50 000 years ago. Whether they brought a dog with them at either of these times has not been established.

However, there is no doubt that the early Australian Aborigine knew of thylacines and coexisted with them. Dr I.M. Crawford of the Western Australian Museum sent me a photograph of an Aboriginal cave painting which clearly shows the features of a thylacine. This painting is from the Kimberleys but Dr Crawford was unable to date the painting to either before or after a cultural change which took place 3000 years ago.

Mrs M. Sack (1981) told me of a tape she made in 1964 when she was the only guest at a farewell party given her by tribal elders from the Marble Bar area. One man, about 70

Aboriginal painting of thylacine, central Kimberleys, 1980; total length, 1·45 m. *Source*: Western Australian Museum and Dr I. Crawford.

years old, from the Pilbara sang 'a song of the dingo but the leader was the one with the stripes', referring to the leader of the dingoes. This particular tribe spent a lot of time in the caves of the Hamersley Ranges before whites came, and this is the only report to come to my notice of an oral thylacine legend in Australian Aboriginal culture. The perpetuity of the legend may indicate that the image was refreshed from time to time by visits to the sacred cave paintings, or that a visit to the paintings was an inspiration to create a new song dating from comparatively recent times. Alternatively, the existence of the song may have nothing whatsoever to do with the paintings and could be a true oral legend dating from the time when thylacines and Aborigines coexisted, perhaps not as long ago as 3000 years.

Brandl (1973) described Aboriginal paintings from sites at Deaf Adder Creek and Cadell River in Arnhem Land. These paintings in monochrome ochre portray animals, some of

which undoubtedly are thylacines. At these sites also are
paintings of dogs and dingoes. The thylacine paintings are
associated with the Mimi art which, according to Aboriginal
legend, was drawn by the spirits or 'old people'. Significantly, the dogs and dingoes were drawn in the X-ray style of the
post-Mimi cultural change period. He was unable to give a
date to the thylacine paintings.

The site at Obiri Rocks, near Oenpelli, has two paintings
of thylacines, one of which is very lifelike. These, like the
paintings from the other sites, are also from the pre-X-ray
period.

Aborigines would have ceased painting thylacines shortly
after the animals' extinction in the area, and the different
technique applied to paintings of dogs and dingoes suggests
that these latter two species did not coexist with the thylacine, at least not in Arnhem Land. This theory would suggest
that Mulvaney's estimate of 8600 years is closer to the mark
than 20 000 years, but even this may be too old a dating.

The continental thylacines may well have suffered predation by Aborigines but it is unlikely that this would be a
major cause of their extinction. It is more probable that
competition with the two canids was the main factor con-
tributing to their elimination from the Australian mainland,
although it is difficult to accept that this competition would
have taken place all over Australia, which is an immense area,
resulting in the complete disappearance of the thylacine from
the continent. Calaby (1971) may have had similar reserva-
tions in mind when he stated that dogs may usurp thylacines
in sympatric areas.

The thylacine was of some importance as food to the
Tasmanian Aborigines (Plomley 1966). The species was
known as *lagunta* or *corinna* to the east coast tribes and as
loarinna to the north-west tribes (Roth 1899). Both the southern tribes and the Bruni Island tribe had two names for the
thylacine, namely *laoonana* and *ka-nunnah* (Milligan 1859).
Thylacines have not been recorded from Bruni Island but the
existence of two names suggests that the animal was well
known to the islanders, who probably encountered it on their
visits to the mainland of Tasmania. Plomley (1966), quoting
the Robinson diaries, noted that thylacines were killed by

Aborigines, apparently whenever possible; Robinson recorded one occasion on which they ate the carcases of three cubs which they had caught. The Aboriginal population of Tasmania was small, probably not exceeding 5000, so this predation could not have been extensive. In Tasmania thylacines and Aborigines lived together into modern times, and since neither dogs and/or dingoes nor Aborigines can be held responsible for the decline and diminution of the thylacine in Tasmania, here we have to look at the activities of whites.

The thylacine and white settlers

The settlement and subsequent exploration of Australia yielded no evidence of the existence of the thylacine on the continent, and the living animal was found only on the island of Tasmania, or Van Diemen's Land as it was then known.

Abel Tasman reported the presence of strange footprints when he landed near Dunalley in 1642, but there is no evidence that they were prints of a thylacine. The first reference was by Paterson when he encountered a fierce animal which may have been a thylacine, and the first definitive identification of a thylacine must be that of Knopwood when he described a 'tyger' on 18 June 1805.

The early white settlers were busy keeping themselves alive and did not have much spare time for writing, and the few records we have now lead us to believe that in the early days thylacines were not numerous and were a matter for comment when seen. For example, Jeffreys (1820) stated that only four had been seen since settlement. These probably would have been the Knopwood and Paterson sightings and the two specimens sent to London by Harris at the time of his original description of the species. More thylacines would have been seen as exploration spread across the country but the point is made that thylacines were not by any means a common species.

Evans (1822), then Surveyor-General of the colony, who lived in Tasmania from 1809 to 1824 and travelled extensively, owning properties in northern Tasmania, commented on the 'opossum hyaena' and stated 'but few of the latter have been seen'.

Widowson (1829) gave some more distributional information when he said of the thylacine that it 'frequents the wilds of Van Diemen's Land and is scarcely heard of in located districts'. Mudie (1829) tells us that the thylacine existed in 'inland Tasmania' but 'did not approach the thickly populated parts of the country'. Both of these reports agree that thylacines lived in the remoter parts of the colony which were largely unexplored and were not used for grazing and farming, and both make the same contradiction when they go on to say that thylacines killed sheep at farms. The presence of farms certainly implies some degree of settlement where the thylacines were killing sheep, and could hardly be called 'the wilds'.

Gunn (1852), who was in Tasmania in the period 1840–60, recorded thylacines 'occurring all over the island from mountain tops to sea level', but Gunn's views of a widespread distribution are not in accordance with those of earlier or even contemporaneous writers. West (1852) at this time went as far as to predict extinction for the species as it had become extremely rare, and Lloyd (1862) noted that the thylacine was 'not very numerous and but seldom seen'.

Gould (1863) said that thylacines inhabited only the tops of mountains, implying that the species had been driven from the more settled areas. By this time there seems to have been some agreement on this point as Gunn (1863) agreed with Gould when he said that thylacines were confined to the remote parts of the state and that they were not usually found in the settled parts of the country. Gould went further and joined West in predicting extinction for the species particularly as sheep had become part of their diet soon after settlement. It was obvious that any thylacine unfortunate enough to be found on a sheep property was not likely to be left alone but would be hunted down and killed.

Clearly, even at this early time, warning bells were being sounded on behalf of the thylacine by a few far-sighted people.

The thylacine and sheep

There is no doubt that thylacines quickly learnt that sheep were easy prey. The first sheep arrived with the initial group of settlers in 1803 and became the basis of vital food and clothing industries . Lycett (1824) complained about an animal of 'the panther kind which commits dreadful havoc among the flocks'. Widowson (1829) protested about the number of lambs that thylacines killed, while Mudie (1829) said that they also raided poultry yards. At George's Bay (east coast) Wedge had trouble with 'tigers killing sheep' in 1831 (Wedge 1962). George Robinson, who walked over much of Tasmania during the years 1830–34, recorded in his diary (17 August 1830) that 'great numbers of hyaenas killed sheep at Surrey Hills' (Plomley 1966).

The conflict between thylacines and pastoralists commenced in the very early days of settlement and was spearheaded by the Van Diemen's Land Company which had vast holdings that included the whole north-western end of the island at Woolnorth (100 000 acres) and also large areas at Middlesex Plains, Hampshire, Surrey Hills, Green Hills, and Burnie. Its main activity was sheep and cattle raising, and the Company soon experienced trouble not only with thylacines but also with Tasmanian devils, wild dogs, Aborigines, and vagabonds. Robinson wrote in his diary on 25 January 1834 that nine or ten dogs and hyaenas were killed at Hampshire in one month and on 4 June of the same year they still were numerous and 'a watchman was employed to look after them' (Plomley 1966).

The Van Diemen's Land Company introduced a bounty scheme in 1830 with the object of ridding itself of dogs, devils, and thylacines. The Company offered:

five shillings for every male hyaena, seven shillings for every female hyaena (with or without young). Half the above prices for male and female devils and wild dogs. When 20 hyaenas have been destroyed the reward for the next 20 will be increased to six shillings and eight shillings respectively and afterwards an additional shilling per head will be made after every seven killed until the reward makes 10 shillings for every male and 12 shillings for every female. A proportionate amount will in like manner be made to the rewards given for devils and wild dogs. (Curr 1830, quoted in S.J. Smith 1980).

The thrust of this bounty was directed mainly at the thylacine, and the increasing scale of the rewards shows that the

Company was determined to get rid of the pests, which obviously were to be found in considerable numbers on its holdings. The Company realized that a major effort would have to be made to reduce the population. The dogs and devils were not regarded as such a serious pest being more numerous and easier to destroy and were worth only half the thylacine rate.

Unhappily, the records of bounty payments from these early days have been lost and there are only two payments of which I have a record, the first of which was seven shillings paid to an assigned servant, McKay, for a bitch hyaena killed at Epping Forest near Hampshire on 22 August 1830, and the second concerned a dead hyaena found by Robinson's party on 29 August 1832 at Mt Cameron. The animal had been killed by natives and it was skinned and the pelt taken to Cape Grim to 'get 10 shillings from the Company'. The scale of reward implies that the Company was paying the maximum amount, so we can conclude that at least forty-seven thylacines had been killed since the inauguration of the scheme in 1830.

The VDL Company in fact had more trouble with wild dogs than with thylacines, Backhouse (1843) telling us that wild dogs were very numerous in the Emu Bay area and that they killed numbers of sheep.

The Company's bounty system must have resulted in financial gain to the employees as the properties supported large numbers of devils as well as thylacines and the bounty was very generous for that time. The reward for thylacines was increased in 1840 to 6 shillings per scalp for fewer than ten scalps, 8 shillings each for ten to twenty scalps, and 10 shillings each for more than twenty scalps, and the sexual discrimination was removed. It is evident that the Company was having much trouble with thylacines, which must have been relatively abundant for such a bounty to be offered.

To combat the high sheep losses on the property at Woolnorth the Company appointed a trapper, who was known as the 'tiger man', and this job persisted until about 1910. He was given a hut and his keep and spent his time trapping and acting as shepherd or assistant as required. He lived at Mt Cameron and visited the homestead every now and again with his skins and dead thylacines, which he exchanged for supplies. His arrival at the homestead was an event which always rated an entry in the station diary.

Later we shall look further into the relationship of the Van Dieman's Land Company with the thylacine (Chapter 6).

We could only make a rough stab in the dark if we tried to estimate how many thylacines were involved in sheep killing and how many of them were killed, or to estimate the thylacine population outside settlements in the wilder areas.

Several trappers, among them T. Pearce (pers. comm.), told me that only a few thylacines became sheep killers, moving down from the forests to kill sheep in adjoining pastures. Some thylacines became addictive killers, Pearce telling me of seeing a thylacine killing sixteen sheep in one night, whilst other trappers spoke of three or four killings in one night. This habit is more typical of the destructive killing by dog packs. On the other hand Pearce was emphatic that many thylacines ignored sheep and would pass through a flock without paying any attention to them.

It is impossible to estimate sheep losses from thylacine killing from the few records we have, particularly when such extravagant claims were made by many farmers, but there is no doubt that sheep losses were sufficiently serious to force farmers to protect their flock against thylacine attacks. A man was appointed at Hampshire Station, and the tiger man at Mt Cameron, Bunce (1857), told of guards being appointed around the midlands flocks at night at lambing time and that at Woolnorth thylacines were often being chased away from flocks.

Hull (1871) and Silver (1874) both commented that the thylacine which had formerly committed such depredations on sheep flocks was now confined to the mountains. The accuracy of these statements is open to question as there is plenty of evidence from the central highlands and the east coast that these depredations were continuing unabated. A property at Blessington allegedly lost 3700 sheep to thylacines between 1865 and 1870, and 448 sheep were killed in the same way on the same property in 1880 (Stevenson 1930, quoted by S.J. Smith 1980). Mr F. Burbury (1953) told me that his property lost 700 sheep out of 2000 at Lake Tooms in one year at about that time. Willett (1927) stated that the losses on the Burbury property at 'The Island' were 450, 350, and 300 in consecutive years.

Other property owners such as Mr Morrison of St Peter's Pass expected 'a hundred or so to be killed' but on the summer pastures of the central highlands 'the losses were much higher but the corpses were eaten by devils and we didn't know how many were killed' (Morrison pers. comm.).

Sheep stealing has always played a part in rural affairs in Tasmania and many of the owners of the large hill runs may not have been aware of the extent of their losses from this source. All of the Pearce family to whom I talked assured me that the losses by sheep stealing were much greater than those sustained from thylacine killings.

On the other hand, not all of the hill runs suffered heavy losses. W. Padman saw nine thylacines in one summer but had only three sheep killed (Padman 1949). Even today, the principal sheep losses in the highlands are due to stealing, and other causes total as low as 1–2 per cent of the losses.

Although it is clear that considerable losses in the sheep flocks were attributed to thylacines many of the deaths may have been due to other causes, especially killing by dogs. Any animal which dies in the bush may be devoured overnight by devils and no evidence of the event is left.

The east coast landholders claimed enormous losses caused by thylacines. J. Lyne of Swansea claimed in 1886 that 30 000 to 40 000 sheep were killed each year on the east coast and that one thylacine would kill 100 sheep per annum. He went on to say that thylacines chased sheep over cliffs and that more were maimed than were killed (J. Lyne 1886). Mr Davies of Fingal (J. Davies 1886) in the same House of Assembly debate declared that 20 per cent of his sheep in mountain country were killed by thylacines. Later in the year Lyne raised his claim to no fewer than 50 000 head of sheep being killed annually by thylacines on the east coast, and he stated that he paid £3 per head for each thylacine carcase. Unfortunately we have no record of how many carcases were presented for payment. This practice would seem to have been fairly general at that time as the Malahide estate at Fingal 'paid 25 shillings per head and has killed about 50'

(J. Lyne 1887). It is illuminating to compare these claims in the light of statements made by other farmers and also in relation to such records as are available to us today.

Sheep losses on the Kelvedon estate near Swansea in the 1885–87 period were about 200 sheep per annum, but most were stolen and only a dozen or so were killed by thylacines, these being taken in the hill country towards Tooms Lake (Cotton pers. comm.). Other landholders in the Swansea area told me a similar story. Mr F. Shaw kept his losses down to about 6 per cent per annum by employing good shepherds and by not using the hill country. This latter statement is interesting as it shows that good husbandry did effectively reduce sheep losses, as was also shown on the VDL Company properties. The implication is that high sheep losses were the result of using hill runs and overstocking and not taking good care of the animals.

The total sheep population of Tasmania in 1887 was about 1.5 million while the number on the east coast was about 115 000, this area being defined for our purpose as extending from Triabunna to St Helens and extending inland as far as the watershed and including Tooms Lake. Allowing a most generous estimate, the number of sheep in the Swansea area was 35 000 (Scott 1965b). The estimate of 50 000 sheep killed by thylacines was nearly 50 per cent of the east coast flocks, a totally untenable situation which would have bankrupted every farmer on the coast. There is no doubt that sheep were killed by thylacines but these claims were grossly exaggerated and losses from other causes were inclined to be attributed to the thylacine.

The persecution and decline of the thylacine

As a direct result of the sheep losses, real and imaginary, and particularly of the claims of the Lyne–Davies group, a petition signed by twenty-six residents of the east coast was presented to state parliament on 28 October 1884 requesting that a bounty be paid on thylacine carcases. A similar petition was lodged on 24 October 1885. These petitions brought no result and the matter appeared before parliament again on 4 November 1886 when the claim of 50 000 sheep lost per annum was made. At that time the rural group was very powerful and, although there was considerable debate, the Lyne motion to pay £1 bounty in the 1887 financial year was carried by twelve votes to eleven.

No other Tasmanian parliamentary action has had such a dreadful effect upon any of the state's fauna. The decision was based upon wildly exaggerated claims which in reality covered up bad farming practice. No attempt was made to check the veracity of the claims, nor was any effort made to ascertain the numbers of thylacines doing the damage. Even a rough calculation based on one kill by a thylacine every three days would have shown that a population of over 400 thylacines would have been necessary to kill 50 000 sheep per annum, and there is no evidence from any source that the east coast area supported this number of thylacines in any one year.

In the course of the debate, some members pointed out that if the losses were as great as claimed then surely the farmers would have done something about it themselves. In actual fact the farmers did do something to help themselves and several stock protection associations were formed at this time. There was a Midlands Association of which H.R. Reynolds was Secretary-Treasurer, an Oatlands Association with F. Burbury as Treasurer, as well as a Glamorgan Stock Protection Association, a Buckland and Spring Bay Tiger Exterminating Association, and a private bounty scheme organized by Hamilton Council. All of these soon faded out and their records went too except for the Hamilton Municipality Scheme, whose records I have been able to examine.

The Minister for Lands and Works, Mr E.N.C. Bradden, announced that the Glamorgan Stock Protection Association would pay $2 for every grown animal and £1 for every pup provided the government would pay rewards equal to those of the Association. The government agreed to do this on 28 May 1888. If a thylacine was submitted for bounty through a stock association it was worth $2 from the association plus £2 from the government. I have not been able to find records of government payments for this purpose and this arrangement

TABLE 2.1   Thylacines killed at Woolnorth, 1874–87

Year	No.	Year	No.
1874	7	1881	9
1875	11	1882	4
1876	5	1883	9
1877	8	1884	0
1878	7	1885	0
1879	1	1886	1
1880	5	1887	3
	Total number killed 1874–87	70	

Source: Guiler 1961a.

presumably was superseded by the government bounty scheme.

At about this time the Van Diemen's Land Company, not wanting to be out of line, increased the rates under its own bounty scheme to a flat rate of 10 shillings for an adult and 5 shillings for a pup, the same as the government scheme. The Company's scheme had continued since 1830 but the records are incomplete and it is only possible to get an approximate estimate of the number of animals killed. Earlier in this chapter we saw that at least forty-seven thylacines had been killed by 1832. A total of seventy were killed between 1874 and 1887 at Woolnorth (Table 2.1), a mean of five per year. This is much lower than the eighteen per year assumed from the estimate derived from the Robinson diaries. A catch of ten thylacines per annum on Woolnorth would not seem to be unreasonable and, if this is extrapolated from 1887 back to 1830, then an estimate of 370 thylacines having been killed during this time may be formed. Unfortunately many of the early records have been lost and it is not possible to trace most of the bounty payments. The cash books from 1841–1885 and the individual accounts from 1849–52 and 1876–1904 show only eight bounties as being paid, two of these being one each to B. Spinks and A. Walker, both of whom were 'tiger men' but may not have been employed as such when they were given the reward. A reward of 10 shillings was a pleasant extra for a family as the monthly wage at the time was about £6.10.0.

The government bounty

The bounty scheme was to commence during the 1887 financial year but due to an oversight the funds from which to pay the bounty were left out of the Estimates for that year. During the Estimates Debate this oversight was pointed out and once again Lyne made his statements about sheep looses.

The sum of £500 was provided in the 1888 Estimates and this was continued until 1907–08 by which time the number of claimants had declined to such a level that a budget item was no longer necessary. Further bounties were paid by a Governor-in-Council's authority. All of the claims were paid through the Lands Department and details are given in Table 2.2.

The procedure was that the successful trapper would take the thylacine carcase to the nearest police station which would then forward the claim to the Lands Department for payment. In many instances the police paid the claims and the Police Department was reimbursed accordingly. A magistrate also was able to issue a certificate and the claim then was sent to the Lands Department. The carcase had its toes clipped off or its ears removed so that it could not be presented more than once for bounty. In some instances the bounty was paid by the Hamilton Municipality through its own scheme and the money recovered from the government scheme.

The first bounty was paid on 28 April 1888 to J. Harding of Ross, and the last to J. Bryant of Hamilton on 5 June 1909. Throughout the bounty period the amount paid was £1 for an adult and 10 shillings for a pup.

Searching the account books of the Lands Department uncovered some basic information. They contained the amount paid to the claimant and whether the payment was for one or more adults or for young, and usually the name of the claimant and the general district in which the claimant lived. By searching the electoral rolls and the postal directories of the period it was often possible to determine the property or area from which the thylacine had been captured.

The number of bounties paid is given in Table 2.2. The numbers given for 1906 and 1907 are based upon the state-

TABLE 2.2   Thylacines presented for government bounty annually,
Tasmania 1888–1912

	Adults	Juveniles	Total
1888	72	9	81
1889	109	4	113
1890	126	2	128
1891	87	3	90
1892	106	6	112
1893	103	4	107
1894	100	5	105
1895	104	5	109
1896	119	2	121
1897	107	13	120
1898	106	2	108
1899	132	11	143
1900	138	15	153
1901	140	11	151
1902	105	14	119
1903	92	4	96
1904	82	16	98
1905	99	12	111
1906	54 (30)	4 (3)	58
1907	42 (19)	0	42
1908	15	2	17
1909	2	0	2
1910	0	0	0
1911	0	0	0
1912	0	0	0
	2040	144	2184

Note: The numbers for 1906 and 1907 were calculated from Treasury reports, the numbers in brackets being the totals for which bounty was claimed through the Lands Department for the first six months of the year. The Lands Department account books for the last six months are missing.

Source: Guiler 1961a.

ments made in the Treasury reports for those years as the Lands Department accounts are not sufficiently detailed and show only the numbers given in parenthesis.

It must be emphasized that this total by no means represents the total kill. Many trappers told me that up to half the thylacines killed were not submitted for bounty but were carted around the local property owners who paid a reward (usually £1) and when the carcase became too smelly it was dumped in the bush. The number of animals killed and used this way is completely untraceable and it is purposeless to speculate further on the topic.

Local folk nowadays say that twelve thylacines were killed at Native Corners at Campania, but I have no information when they were killed or whether they were used for bounty or otherwise.

Years ago I talked to old-timers who obviously had had experience of trapping thylacines. Their names do not appear on the bounty lists, but the animals may have been taken around properties or submitted to one of the stock protection organizations. The late Mr F. Burbury of Parattah when treasurer of the Oatlands–Ross Landowners Association paid shepherds a bounty of £5.10.0 for each carcase, and he recalled that about forty bounties were paid—one shepherd collecting on eighteen thylacines (Burbury 1953).

Crowther (1883) while reassuring us that the thylacine was not extinct wrote that 'hawkers from the interior' frequently offered skins for sale in Hobart Town. These pelts may well have come from those trapped by snarers who set gin traps around their possum and wallaby snare lines; thylacines and devils were caught in this way especially in the important fur-trapping areas of the central highlands. These snarers may also have been responsible for the trade in thylacine skins selling for £3.18.0 each when tanned. No fewer than 3482 skins were exported between 1878 and 1896 (Laird 1968). These pelts would have come through the normal fur trade channels. According to Laird and Mollison (1951) these skins were in demand for waistcoats. Unfortunately I have not been able to trace Laird's sources. He died some years ago and his papers do not contain details of this interesting material. In addition there would have been many thylacines snared whose skins were so damaged as to be valueless and these would have been dumped in the bush.

Throughout the period of the government bounty the Van Diemen's Land Company continued with its own scheme, and its employees made claim on the company rather than on the government. There are a number of entries in the individual accounts for bounties paid to employees during this period and some of them were 'tiger men'. I have not in-

TABLE 2.3   Thylacines killed at Woolnorth, 1888–1914

Year	No.	Year	No.
1888	3	1901	9
1889	2	1902	3
1890	6	1903	2
1891	7	1904	0
1892	2	1905	0
1893	5	1906	1
1894	4	1907	0
1895	6	1908	0
1896	3	1909	0
1897	4	1910	0
1898	2	1911	0
1899	3	1912	0
1900	19	1913	0
		1914	3

Total number killed 1888–1914   84

Source: Guiler 1961a.

cluded these in the calculations to obtain the total number of
thylacines killed under bounty as I believe that the Company
ultimately made claim on the government scheme.

The number of thylacines killed during government bounty times on Woolnorth was eighty-four, with the year 1900
yielding no less than nineteen, but this was the peak and by
1903 there were only two captures (Table 2.3).

The catch of thylacines under the bounty scheme was
steady at about 100 per annum until 1905 when the number
fell dramatically, reaching zero in 1910 (Table 2.2). The
Woolnorth catches declined in the same fashion. This final
decline, which was very rapid and occurred all over the state
at about the same time, is not typical for a species which has
been hunted to extinction.

In the last year of the bounty, thylacines were caught at
Bicheno and Cranbrook on the east coast, Hamilton on the
edge of the central plateau, Mike's Hill near New Norfolk
and Broadmarsh, the latter two places being within 35 kilometres of Hobart.

If the thylacine had been hunted to extinction, it is my
view that it would be logical to expect the animals to disappear first from the places where they had been most vigor-

ously hunted since early settlement, but this did not take place. The last thylacine caught on Woolnorth was taken in 1914 after the species had suffered nearly 100 years of persecution, and thylacines had been hunted for many years at Bicheno and Cranbrook on the east coast, and in the Hamilton area which had active graziers' societies dedicated to eliminating thylacines and other pests.

Similar declines can be seen in other species—in the case of the hunting of the blue whale the catch peaked in 1930–31, then declined until 1940 when the war gave a respite, and they were still caught in numbers until 1960–61 when they faced extinction. This decline extended over thirty years, with no sudden decline as seen in the thylacine population. No other species has declined so rapidly, probably too rapidly for hunting to be the sole cause, and we have to look for other factors which may have been partly responsible.

Thylacine habitat has been drastically altered since the arrival of Europeans in Tasmania. The alteration took place quickly and by 1844 large areas of the midlands, the upper Derwent Valley, and the east coast had been granted or sold to settlers, quite apart from the large Van Diemen's Land Company holdings. By as early as 1831 there was little woodland left for alienation. This expansion lasted until 1914, after which areas began to revert to the crown.

Although land was alienated it does not necessarily mean that it was altered in any way other than by having sheep or cattle grazing on it and by being cleared in valleys and some fencing erected. However it was subject to some change and to human interference, and this could have had a disturbing effect not only on thylacines but on their prey as well.

The south west remained largely untouched by the land developments until the advent of the Hydroelectric Commission dam projects, forming Lake Pedder and Lake Gordon, which flooded potentially useful thylacine habitat. In addition, logging took place over some of this country, particularly near Adamsfield. However much of the south west is still undeveloped, it is not used for pastoral activities, and the important coastal strip remains untouched (Scott 1965a).

A clue to a possible factor in the decline comes from F. Burbury (1953) who wrote:

about 1910 the thylacines seemed to disappear, also the indigenous cats, native cat and tiger cat, seemed to disappear. I know a form of distemper killed the cats but I have no practical evidence that a similar cause killed tigers. The real fact was, the destruction of sheep ceased, and others with me thought it was because of distemper.

Hickman (1955), Skemp (1958), and Lester (1983) all mentioned the possibility of a distemper-like disease, and mange was mentioned by some (S.J. Smith 1980).

Little is known of the diseases of thylacines or, indeed, of many of the other marsupials (Chapter 5). A form of pleuropneumonia is a frequently encountered disease in captive colonies of the potoroo, *Potorous tridactylus*, and it is known that large numbers of the ringtail possum, *Pseudochirus convolutor*, died of this complaint during a violent population crash in this species in 1952–53 (Guiler 1967, 1971). However, it is clear that distemper, which is a disease of canids, could not infect thylacines, but a disease having similar symptoms might have contributed to their sudden demise all over Tasmania.

The very rapid decline of the thylacine was not due to any one factor but was the outcome of a combination of all of the above factors. The habitat alteration and consequent disturbance prevented thylacines from operating in their favoured areas and could well have caused territorial disputes between individuals. This was combined with continual persecution which reached a peak at a time when disease chose to strike the Dasyures as a whole. All of these factors combined to depress the numbers below a satisfactory breeding threshold and so prevented their recovery from what ought to have been a normal cyclical change. It should be noted that native cats have recovered their population numbers and now are common again over much of Tasmania.

There may have been competition with wild dogs during the period 1820–1900; there is abundant evidence in the station diaries that dogs killed many sheep on VDL Company properties and it could be expected that this happened elsewhere throughout Tasmania. If the presence of dogs on continental Australia was a factor in the extinction of the thylacine, then it cannot be eliminated from the Tasmanian situation.

Whatever the causes of the decline, the reduction of the thylacine population to near-extinction level was achieved very cheaply from the point of view of the government, costing an average of only £110 per annum for twenty years. This must surely be the cheapest extermination campaign in the sad history of predator control. Those twelve parliamentarians who voted for the Bill to pay a bounty without inquiring into the authenticity of the facts and figures given have a great deal to answer for in the history of conservation in Tasmania, and the species may never now recover from the effects of that campaign. Extinction for the thylacine was predicted by some early writers such as West (1852) and Gould (1863), and echoed by other more recent authors, notably Flynn (1914), Renshaw (1938), Le Souef (1926), Vrydagh *et al.* (1964), Archer (1978), and Rounsevell (1983) who went so far as to suggest that the species was extinct by 1936.

The last thylacine in capitivity died in the Hobart Zoo in 1936, while the last one in an overseas zoo died in London in 1931. The last confirmed killing of a wild thylacine was in 1930.

Legal protective measures

Although it must have been apparent to all by about 1920 that the thylacine was very rare and a highly endangered species, nothing was done to afford it any protection. There does not appear to have been any pressure exerted upon the government by any public body or vigilante organization to do anything about it. In fact, snaring thylacines for zoos still occurred well into the 1930s.

It was not until 20 August 1929 that a motion was passed by the Animals and Birds Protection Board, the statutory authority charged with responsibility for faunal matters, to provide for the partial protection of thylacines by closing the open season for the month of December. The fact that there was an open season does not imply that people were encouraged to buy a licence and go off to hunt thylacines. Under the regulations of the day, any species which was not protected, either wholly or partially, was subject to a twelve-month open season. A closed season was gazetted on 6 May 1930, prohibiting the hunting of thylacines during December, the rationale being that this month was believed to be the breeding season.

A subcommittee was established on 19 September 1933 to consider measures to protect the species but little appears to have come from its deliberations as permits to catch thylacines for Australian zoos continued to be issued up to 1936.

The partial protection did not produce the hoped-for increase in numbers and consequently, after concern for the species had been expressed at several meetings of the Board, the thylacine was declared a totally protected species on 14 July 1936, a status which it still holds today.

At the time of the gazettal of the protection notice, a suggestion was made that the surviving thylacines should be rounded up and placed on De Witt Island, a miserable place off the south coast which has no suitable habitat and very little prey for thylacines. Fortunately, nobody caught any thylacines and they were spared this further trial.

It is somewhat ironic that by the time the thylacine was being granted total protection the first of the major thylacine searches was being contemplated, the expedition finally leaving Hobart in November 1938. The government action was far behind the needs of the thylacine and indeed the search expedition was equally belated.

Sanctuaries

Various approaches were made from time to time to the government suggesting that attempts should be made to capture thylacines and then to rehabilitate the species in some sort of sanctuary. These suggestions started when the then Ralston Professor of Biology at the University of Tasmania, Prof. T.T. Flynn, foresaw the future scarcity of the species and suggested that some be captured and placed on an island (Flynn 1914). Summers (1937) and Sharland (1939), both of whom were involved in early searches for thylacines, recommended that sanctuary areas should be established between the Arthur and Pieman Rivers and in the Jane River area respectively. Unhappily, these suggestions were not taken up at the time and there still is no reserve in that area. The Arthur to Pieman

River proposal could have provided recreational facilities nowadays in an area not well served by parks.

A sanctuary for thylacines would not be easy to establish. It would have to contain sufficient abundance of prey animals in a habitat suitable for both prey and thylacines, the area would have to be of sufficient size to allow movement for the thylacines and provide resting sites and yet prevent 'overflow' of the predators and prey to neighbouring properties—not an easy arrangement to achieve.

Many of the present state reserves do not fill these conditions. The larger parks such as the South West National Park and Cradle Mountain National Park offer enough area for thylacines but only the fringes and the southern end of Cradle Mountain Park offer a suitable habitat. The Freycinet Peninsula National Park and, to a much lesser degree, the Rocky Cape Park might offer some habitat for thylacines. Ben Lomond, Mount Field, and Frenchman's Cap are all large parks of sufficient size and in those places where the habitat is suitable they might well offer good sanctuary for thylacines. The protection of thylacines, or any other species for that matter, was not a consideration in establishing any of these parks, indeed prior to 1950 the principle guiding the selection of parks seemed to be: if the scenery is good and nobody else wants it, then it could be a park.

The notion of a reserve for thylacines seems to have been little pressed after the Summers and Sharland suggestions. The animal had become so rare, if not extinct, that the proposition was not considered to be worth pushing. It was not until the 1963 Animals and Birds Protection Board (Fauna Board) expedition to capture a thylacine took place that the question of a sanctuary came to the fore, largely in the context of what to do with a thylacine if one was caught. The Fauna Board searched around for a suitable place and several areas were considered, particularly the Arthur–Pieman River suggestion of Summers, and Rocky Cape, Tooms Lake, and the Freycinet Peninsula.

The Arthur River area was discarded because it was subject to mining and forestry leases, and Rocky Cape was too small; and both areas presented problems in fencing to retain the animals in the sanctuary. The Freycinet proposal, which was

too small, envisaged the extension of the existing park to include all the land as far north as the top of Moulting Lagoon
and fencing across the 3 miles (4.8 km) or so of neck. This
was ruled out because of the difficulties associated with Coles
Bay township right in the middle of the area with problems
of access for residents, of holidaymakers flocking in and
walking all over the Peninsula, and of excluding dogs and
cats and other undesirable pets from the township area. So
much logging took place around the Tooms Lake area that it
was discarded as not suitable.

As well as an area to accommodate thylacines the Fauna
Board was anxious to set up a sanctuary where the public
could see and appreciate our native animals in their wild
state. The area had to be located so that animals moving out
would not cause conflict on neighbouring properties by eating the pastures. This can be partially controlled with fencing
but is very expensive and not by any means fully effective. A
further financial possibility was to find land already in the
hands of the crown or at least able to be acquired at a modest
cost.

The Fauna Board eventually decided, in 1967, upon Maria
Island, off the east coast, as fulfilling most of the criteria. It
did not need to be fenced. There was some private land on
the island but the cost of resumption was not great as island
farming was becoming uneconomic. Above all, it was large
enough to support a large wallaby and kangaroo population
which could, in turn, support thylacines. Since then the
National Parks and Wildlife Service has superseded the Fauna
Board and controls the management of the island. Forester
kangaroo have been introduced and are now well established,
and there are abundant wallabies and pademelons. The island
has become a very successful wildlife sanctuary and is a
popular recreational resort.

3
The animal

It is not my intention to give a detailed description of all of the anatomical features of the thylacine. Indeed, I shall confine this chapter to descriptions and discussions of those parts having some bearing on problems which arise later in the book. Although there is some knowledge of the anatomy of the thylacine, there is very little known of its physiology.

Knopwood noted in 1805 that a 'tyger' had been seen, about the same time as the possible sighting by Paterson, but three years passed before the species appeared in the scientific literature when a brief description of the animal was given by Harris (1808) who gave it the name *Didelphis cynocephalus*. Harris was the first Surveyor-General of the colony of Van Diemen's Land and presumably as well as seeing dead thylacines he also saw living ones during his work in the bush. The delay of three years from its first recorded sighting until its description in the zoological literature is understandable with communications so slow in those early days. Indeed, it is remarkable that Harris had the time to indulge in describing new species of animals since his surveying duties must have been enormous.

TABLE 3.1   Synonomy of the genus *Thylacinus*

Genus	Species	Synonyms	Author
Thylacinus			Temminck 1824
		Peracyon	Gray 1825
		Lycaon	Wagler 1830
		Paracyon	Griffith, Smith & Pidgeon 1827
		Peralopex	Gloger 1841
	cynocephalus		Harris 1808; Fisher 1829
		Didelphis cynocephala	Harris 1808
		Dasyurus cynocephalus	Geoffroy 1810
		Thylacinus harrisii	Temminck 1824
		Dasyurus lucocephalus	Grant 1831
		Thylacinus striatus	Warlow 1833
		spelaeus	Owen 1845
		breviceps	Krefft 1868
		major	Owen 1877
		rostralis	de Vis 1894
	potens		Woodburne 1967

Sources: Iredale & Troughton 1934; Ride 1964; J.W.J. Lowry 1972.

A French naturalist, Geoffroy (1810) turned his attentions to the marsupials and realized that the genus *Didelphis*, to which Harris had assigned the thylacine, was not suited to the species and he placed it in *Dasyurus* which he had erected to cope with the flow of new species of carnivorous marsupials beginning to arrive from the new continent.

The genus *Thylacinus* was erected by Temminck (1824) when he moved the species out of *Dasyurus* because he believed that it was sufficiently different to warrant this change. He also gave it a new minor name, *harrisii*, which was not accepted. The species has been quite clearly defined and, apart from some fussing about in the early times, has suffered little from the attentions of taxonomists. Some new names were proposed but all have been dropped under the rules of priority. A full list of the synonymy is given in Table 3.1.

The species *T. spelaeus*, *T. major*, and *T. rostratus* all are based upon fossil fragments from continental cave deposits and these were considered by Stephenson (1963) to be synonyms of *T. cynocephalus*. One fossil species, *T. potens* Woodburne (1967), still is regarded as a valid species.

Krefft (1868) erected a new species to cater for what he believed to be a new species of thylacine having a shorter skull than *T. cynocephalus*, which he described as being only 6⅞ inches (175 mm) long with smaller palatal apertures and with a larger occipital foramen. Also, there was no notch between the second and third premolar teeth, and the second and third molars in each jaw were prominent and the canines were thicker than in *T. cynocephalus*. He went on to name this species *T. breviceps* from the short skull and called it the 'bulldog tiger'; he rechristened *T. cynocephalus* as the 'greyhound tiger'.

Allport (1868a) very soon corrected Krefft's views and pointed out that the so-called bulldog tiger with the short skull was in fact the female thylacine whilst the greyhound tiger was the male of the species. Krefft had in effect erected a new species to describe the sexual dimorphism of *T. cynocephalus*. Very much later, Moeller (1968) showed that Krefft's skull was that of a young animal.

The common names bulldog and greyhound tiger, referring to the female and the male, were used in the central highlands and they were known elsewhere. The source locality of Krefft's specimen was Ouse where these names would have been in common usage. It may be significant that the hotel in Ouse is well known for the proficiency of some of its habitues in the honoured pastime of pulling the leg of any stranger and perhaps Krefft fell for one of their efforts.

The most frequently used vernacular name is undoubtedly 'Tasmanian tiger' or more simply 'tiger', but many other names have been used over the years, some of which were never in common use. The terms 'wolf' or 'Tasmanian wolf', 'hyaena' and, to a lesser extent, 'thylacine' were used, whereas such names as 'dog-faced dasyurus' (Mudie 1829), 'dog-headed opossum' (Angas 1862), 'opossum hyaena' (Evans 1822) or Harris's 'zebra opossum' would not have been commonly known and more likely would have been used in the literature rather than by the bush folk. Combinations of terms were used such as 'marsupial wolf' (Lucas & LeSouef

1909), 'striped wolf' (Wright 1892) and 'Van Diemen's Land tiger' (Henderson 1832). The name 'dingo' was used infrequently (J. Lyne 1886) and this was not only unusual but incorrect. As far as I can determine the term 'thylacine' was first used by Krefft in 1868. I never heard the old-timers refer to the animal as anything but 'tiger' or 'hyaena', or more rarely 'wolf'. The geographical prefix was not used by them. The most widespread current name is 'tiger' or 'Tasmanian tiger', but as 'tiger' infers a relationship with the cats I shall not use 'Tasmanian tiger', though 'tiger' will be used from time to time, especially in relating stories by old-timers.

To this list of common names we should add 'slut' as referring to the female of the species. I am not sure how widely this term was used. It certainly was in general use in the central highlands and by the Pearce family always when speaking of the female.

Relationships

The thylacine is considered usually to be in a family of its own, the Thylacinidae of the order Dasyuroidea. This order of marsupials was established to include all of the carnivorous marsupials of Australian origin and includes such animals as marsupial mice and phascogales and their larger relatives, the native cats, tiger cats, and Tasmanian devils.

It has been suggested that the Thylacinidae of the early evolutionary period in the Miocene were quite distinct from the Dasyures (Woodburne 1967). A similar suggestion had been made by Bensley (1903) who postulated that *Thylacinus* was 'an unrelated element in the Australian family'. Archer (1976a) revived this notion and, basing his evidence upon tooth and skull structure, suggested that the genus was related to the fossil Borhyaenids of South America, another carnivorous group of marsupials. However, in a later paper Archer (1976b) found that there was a relationship with the South American opossums rather than with the Borhyaenids. Ride (1964) showed that there were differences between *Thylacinus* and the other Dasyures but Simpson (1941) and Tate (1947) preferred to regard it as a Dasyuroid.

This was the slightly confused situation up to the development of serological studies which showed that the relationship is closer to the Dasyurids than to any other group and that the closest relative to *Thylacinus* is the Tasmanian devil, *Sarcophilus harrisii* (Kirsch & Archer 1982; Sarich *et al.* 1982; Archer 1982).

External appearance of the animal

The thylacine shows a remarkable similarity in form with the more widely known wolf and dog, having the typical form of large head, evenly balanced backbone, legs of approximately equal length, and hind legs with the femoral segment pointing obliquely forwards. Like the dog, it has the deep chest and non-retractile claws, and runs on its toes and not on the whole foot. However, in spite of these convergences in external form the thylacine remains in all of its features very clearly a member of the marsupials.

There are scanty measurements of the thylacine and even fewer of its weight. Much of this information was collated by Moeller (1968) who used museum specimens as well as sources from the old literature such as Harris (1808), Breton (1834, 1847), and Crisp (1855).

The head and body length ranges from 851 to 1181 mm (mean = 1086 mm), tail length 331 to 610 mm (mean = 534 mm), height at the shoulder about 560 mm, the whole weighing about 25 kg. Several of the old-timers made the point that thylacines frequently were about 7 feet in length (2133 mm) and Sawley (1980) claimed that a 'tiger shot at McKay's property at Trowutta was the biggest ever known in this area measuring 9 feet [2716 mm] from the tip of nose to tip of tail'. It is possible that larger animals existed than were ever sent to museums, universities, and zoos.

The most characteristic feature of the thylacine is the 13–19 dark-brown coloured bands extending from the posterior thoracic region onto the butt of the tail. These bands provide contrast to the brown, light brown, or sandy colour of the body. The more anterior bands extend only a short distance from the midline whilst the longest bands occurring on the rump extend laterally as far as the upper thigh. The posterior rump bands are short and extend onto the butt of the tail.

The number of markings varies from individual to individual and the coloured bands are more prominent in the younger animals.

Renshaw (1938) noted that this disruptive colour pattern is rare amongst mammals and listed only two species which share this feature with the thylacine. In this he is not strictly correct as there are more than two species, but the colour pattern is not common and is found usually on mammals which live in savannah woodland or forest areas. The barred bandicoot, *Perameles* spp., uses this pattern to advantage in the light and shade of the tussock grasses where it lives, while the feline tiger and the okapi have similar colour patterns living in dense forests. It can be concluded that the thylacine with its disruptive colour pattern is adapted to living in woodland conditions where it can gain camouflage in light and shade from its colouration.

The body hair is short and dense and usually fawn to yellow-brown, though some individuals, usually young, are a darker brown shade. The ventral belly fur is light in colour and of a creamy hue.

The tail does not have a canine appearance and, unlike that of a dog, cannot be wagged laterally. Thus the thylacine when turning gives the impression of being clumsy and ungainly. The fur on the tail is close and tight which is in contrast to the often longish fur of a dog. The continuation of the body stripes onto the thick butt of the tail gives the impression that the tail is a continuation of the body, whereas the tail of a dog is clearly delineated from the trunk.

The female has a backward-opening pouch which is nearly circular and contains four nipples. A forward-opening pouch could be a liability for rapidly moving ground-dwelling marsupials as grass or twigs could easily penetrate the aperture whereas a backward-directed aperture prevents this. Other ground dwellers such as bandicoots and Tasmanian devils also have backward-opening pouches.

The male thylacine is unique among marsupials in that the testes, which in the marsupials are external and unprotected except by their inguinal position, are carried in a partial pouch (Beddard 1891; Pocock 1926; Hickman 1955). The testes are given greater protection than can be given by the normal method of scrotal retraction. The male Tasmanian devil represents an intermediate stage in that the testes are protected by the lateral folds of the belly skin which form a 'pseudo-pouch' offering considerable protection to the testes (Guiler & Heddle 1970).

More details of the externals are to be found in the paper by Pocock (1926).

Posture
In his description of the anatomy of the thylacine, Cunningham (1882) devoted much time to the musculature and makes the point that the limb muscles are typical whereas those of the dog are atypical in that the flexores breves and adductors of the pes are coalesced. The flexores breves provide for continuous and maximum flexation of the digits without interfering with the wrist movement. It may be inferred that the dog requires rigidity of the wrist and digits to enable it to move rapidly, while the thylacine does not have this adaptation and is a more slowly moving species. Cunningham went further and said that the thylacine had a slow skulking habit. This suggestion is supported by those persons fortunate enough to have seen thylacines hunting.

The remarks of Cunningham are particularly pertinent when considered in relation to gait and posture. It has been stated that the thylacine adopts a bipedal kangaroo-like posture and can leap like a kangaroo (Mattingley 1946; Moeller 1968) but this is not supported by any evidence in Cunningham's study. Indeed he states that both gait and posture are typical of a quadruped.

Barnett (1970) examined the ankle joint in some detail and found no unusual features in this joint, and Cave (1968) in his study of the olecranon process of the elbow did not find any features incompatible with quadrupedal locomotion. We are left with the impression that the skeleton of *Thylacinus* shows no adaptations for bipedal activities.

However, there are some skeletal differences from other marsupials, notably there being only two sacral vertebrae (compared to the usual three) and the normal 20–21 caudal vertebrae are replaced by 23–25. The increase in caudal vertebrae might be expected in view of the length and weight of

Fig. 3.1   Head of a thylacine showing the vibrissae. Based on a drawing by Pocock (1926) but modified according to a specimen in my own collection.

the tail (Crisp 1855; Moeller 1968). Some elements of the skeleton are reduced, notably the clavicles (Cunningham 1882) and, surprisingly, the marsupial bones of the pelvic girdle which are rudimentary and unossified (Owen 1843, 1846). The marsupial bones are used as a support for the pouch in other marsupials and it might be expected that an active predatory thylacine carrying four young in its pouch would require more support to help carry the young during hunting. There is nothing to suggest that the pouch young quitted the pouch at an early stage thus reducing the load upon the pouch and its supports. The reason for the rudimentary nature of these bones must be a matter for speculation.

Vibrissae and hairs
The vibrissae are the strong whiskers found usually in the facial region of mammals though they do occur elsewhere on the body. The vibrissae of marsupials were described by

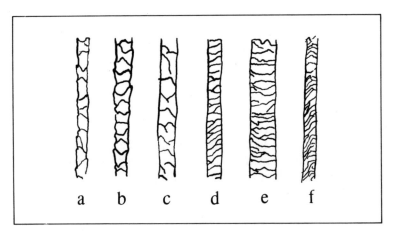

Fig. 3.2 Surface structure of the hair of *Thylacinus* (from A.G. Lyne & McMahon 1951). **a, b** basal part of a wavy hair; **c** mid-shaft of wavy hair; **d, f** distal portion of wavy hair; **e** distal portion of straight hair.

Pocock (1914) and Wood Jones (1929), and comprehensively reviewed by A.G. Lyne (1959), who quoted the marsupial mouse as a primitive example from the Dasyures, having a full set of vibrissae. The thylacine is regarded as specialized and has fewer vibrissae, the only representatives being the facials and supraorbitals together with the interramal between the rami of the lower jaw and the submental on the lower lip. All of the supraorbital vibrissae are very long as are the mystacials, especially those of the upper row (Fig. 3.1).

Every species of mammal has its own characteristic structure of the hairs, the guard hairs having a different structure from those of the body. Two features of the hairs can be used to identify the species from which the hair was derived, namely the cross-sectional shape and the external cuticular scale pattern. A.G. Lyne & McMahon (1951) described the hair patterns for marsupials and developed a simple technique for examination of the cuticular scale pattern using a film emulsion-nigrescin stain method. The cuticular scale pattern for the thylacine is shown in Fig. 3.2. By the use of this method it is possible to determine whether the hair in question is that of a thylacine.

Head and skull
The head of the thylacine is remarkably like that of a dog although the ears are shorter and rounder, but otherwise they are similar.

The features of the teeth and skull enable us to identify the skull with some ease. Teeth have fascinated some zoologists and those of the thylacine are no exception, having been described by Owen (1841), Thomas (1888), and Bensley (1903) among others, whilst Moeller (1968) gave a very detailed description of the dentition. It is not my intention to describe the teeth other than to note that they are typically carnivorous in form and adapted for the shearing of meat. The only milk dentition which has been described was a deciduous premolar tooth found by W.H. Flower (1867). The dental formula is $4/3$; $1/1$; $3/4$; $4/4$. The unequal number of incisors in the upper and lower jaws is one of the features which is different from the dog.

The skull also exhibits the following characteristic marsupial features:
1. The lachrymal foramen opens on the anterior surface of the lachrymal bone at the anterior of the orbit and not on its inner surface as in the eutherian mammals.
2. There are two large foramina on the palate of the skull whereas eutherian mammals have none, having a complete palate.
3. The lower jaw has a reflected angle at the inner posterior surface.

The only marsupial skulls with which the thylacine's might be confused are those of the wombat or the Tasmanian devil, all the other species being smaller in size or having obvious kangaroo features. The wombat skull is massive and has a very wide gap between the incisor and molar teeth which is typical of herbivores. The Tasmanian devil has a squat, powerful skull with a short face and all the teeth closely placed together, and with a dental formula of $4/3$; $1/1$; $2/2$; $4/4$. Further details of the skull of the thylacine can be found in Thomas (1888), Tate (1947), Moeller (1968) and Archer (1976b).

The skull serves as an attachment for the muscles associated with the lower jaws and with the connection of the skull

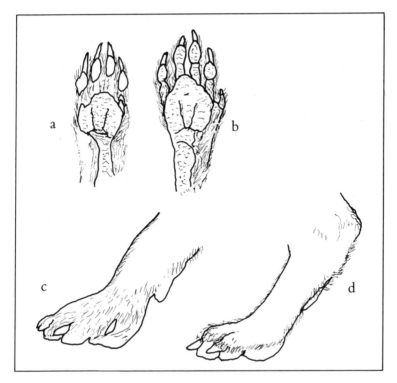

Fig. 3.3   Drawings of the feet of *Thylacinus* based on dried specimens from the Tasmanian Museum. **a** right hindfoot; **b** right forefoot; **c** and **d** right forefoot and hindfoot in normal walking position.

to the neck. These surfaces are large, especially those used for the chewing muscles.

Feet and footprints
The footprints of thylacines are of prime importance in any evaluation of the field evidence for the presence, or otherwise, of the species. Unfortunately, the similarities with the dogs that we have seen so far are equally evident in the feet.

The features to be examined in any spoor are the size of the pad and its shape, the number of toes and the shape and size of the toe pads, the number of claws and their size and whether they appear in a spoor, the spacing between the toe

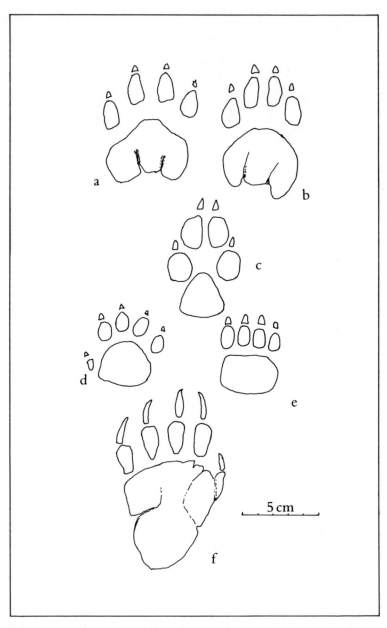

Fig. 3.4   Footprints of thylacine, dog, devil and wombat. **a** left front foot of thylacine; **b** left rear foot of thylacine; **c** dog; **d** right front foot of devil; **e** rear foot of devil; **f** left front foot of wombat.

pads as well as the foot pads, and finally the distance between each step and the nature of the gait.

The feet of *Thylacinus* are shown in Fig. 3.3. They were originally described in great detail by Pocock (1914). The plantar surface of the forefoot of the thylacine is separated from the toe pads by a prominent space seen in the spoor (Fig. 3.4). The toes are arranged symmetrically around the rather irregularly shaped foot pad (Fig. 3.3). The latter has two deep grooves extending forwards from its rear border. The toe pads are of moderate size. The figure given by Pocock is somewhat misleading in that it shows five toes on the front foot. In actual fact, one toe is raised above the others so that in all but soft mud this toe will leave little impression. The arrangement of the toes on the manus is such that numbers three and four are of equal size and placed at equal distances in advance of the foot whilst number two and five toes are placed at the same level as each other but further behind the other two toes.

The plantar surface of the pes is easier to identify in the field because the space separating the digits and main pad is larger, and there is only one plantar groove which is prominent running forwards from the posterior border of the foot. The toes are arranged in a similar pattern to those of the pes except that there are only four of them. This spacing of the toes is very prominent in the spoor. The claws of both feet are to be seen in the spoor but they are not prominent.

The tarsal segment of the leg bears a granulated surface and may be applied to the ground on occasions. This does not appear in the spoor except under unusual circumstances. For example, I saw this segment in a spoor found in a tractor track, and the hock of the animal had left an impression in the steep muddy side of the tyre marks.

It is essential to note that the spoor of the fore and hind feet are different whereas in the dog they closely resemble each other. It is important to have a good set of prints, preferably four, to identify both front and rear spoor. Prints should be examined *in situ*, as digging up the mud or clay in which the spoor is found nearly always results in some distortion of the impression, especially when the mud dries out and cracks.

The distance between each subsequent print of any one foot in an adult thylacine should be about 80 cm or more.

Thylacine spoor filled with plaster, Woolnorth, May 1960.

This is important in eliminating certain other species from consideration in the identification. There are two species whose spoor can be confused with that of the thylacine. These are the wombat and the dog. The spoor of the Tasmanian devil should never offer any problem in this regard since devil pads are small and have a characteristic forefoot which leaves an almost square impression with the toes arranged neatly in an equally spaced row along the front of it.

Wombat spoor can be identified by the large size of the plantar pad of the manus which has large toe pads bearing prominent claws which are strongly curved. All five of the claws appear in a spoor. The trail of a wombat is shuffling and wanders about from bush to bush whereas that of a thylacine is more direct, travelling in a definite direction.

Dog spoor is not easy to distinguish from that of the thylacine and both species travel in the same fashion and have the same gait. The plantar surface of the dog is always subtriangular with the apex pointing towards the front. The median pair of toes is some distance in front of the two lateral toes giving a 'two up, two back' appearance. The toe pads are

TABLE 3.2   Guide to spoor identification

	Thylacine	Dog	Wombat
Forefoot			
Main pad	Irregular, ovoid	Subtriangular, apex anterior	Irregularly subtriangular, apex posterior
Toes	5 one usually does not give an impression	4	5
Digital pad	Moderate size	Large, oval	Moderate
Claws	Small relative to digital pad, narrow space between claw and digit and between digit and main pad	Large, large space between digit and main pad	Large, curved
Hindfoot			
Main pad	Subovoid	Subtriangular, apex anterior	Oblong
Toes	4	4	5 very 'hen-toed' in rear foot, first digit reduced

large and the median toes are at some distance from the plantar pad. The claws on all four toes are larger than those of *Thylacinus*. The forefoot of the dog has the same spoor as the rear foot and this is readily seen in a full set of footprints. The main features of the spoor of thylacine, dog, and wombat are set out in Table 3.2. No indicaton of the size of the spoor has been given as that is a poor guide since it merely indicates the set out in Table 3.2. No indication of the size of the spoor has as those of a moderate-sized dog and dog spoor no larger than those of a big cat. However, in general a print larger than 4.5 cm in width is likely to be that of a wombat, a dog, or a thylacine.

Internal structures

Although we are concerned mainly with the ecology of the thylacine we can make some deductions about its habits from

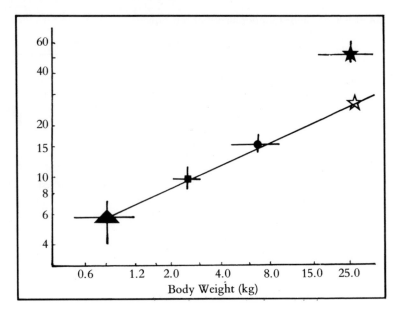

Fig. 3.5   Body weight/brain cavity capacity in *Dasyurus* (triangle), *Dasyurops* (square), *Sarcophilus* (circle), and *Thylacinus* (solid star). The unshaded star is the point at which *Thylacinus* would be expected to occur if the brain-size/body-weight relationship was the same as that for the other three genera. From Moeller, 1970.

the already existing descriptions of the anatomy of the soft parts.

The brain

The brain was described by Owen (1837), W.H. Flower (1865), Beddard (1891) and more recently in a comparative fashion by Moeller (1968, 1970). The brain of *Thylacinus* is much larger than that of a native cat or Tasmanian devil. This might be expected but Moeller goes on to point out that this increase in brain capacity is much greater than would be anticipated from an allometric shifting of the brain capacities of the native cat or Tasmanian devil (Fig. 3.5). Not only is the brain larger than in the other two species but it is also relatively larger which may lead us to conclude that the brain is more developed in *Thylacinus*.

Moeller (1970) proceeds to show that the greater development of the brain has taken place in the neocortex but then

points out that the olfactory lobes are reduced and comparatively much smaller than in *Sarcophilus*. The surface of the
brain has many more dorsal furrows than are to be found in
the other three species examined by Moeller (1970).

The findings by Moeller are important to us in trying to
gain an insight into the behaviour of thylacines. It has been
shown that the Tasmanian devil is a scavenger (Buchmann &
Guiler 1977) and the large olfactory organs are related to the
importance of this sense in locating food. The brain structure
of the thylacine shows that this sense is not so important in its
hunting techniques. The thylacine is a larger animal and can
see over the top of grass and low bushes and it can be expected that the sense of smell would be replaced by vision
and sound as the more important hunting senses. The ridges
on the surface of the neocortex are associated with increased
intelligence and this would be required in the hunting role
thylacines have adopted to catch their prey.

Alimentary canal
The alimentary tract was described by Crisp (1855), who
noted that there was no caecum and that the canal was short,
being only 6 feet 6 inches (approx. 200 cm) in length in a
male with a head and body length of 3 feet 3 inches (approx.
100 cm). This is unusually short. He noted also that the
stomach was very muscular and capable of great distension.
This latter observation is of interest to us as it points to a
feeding habit involving the input of a large amount of food,
probably at irregular intervals, the animal eating its fill
whenever the opportunity presented itself. This type of feeding behaviour is in keeping with that of most carnivores and
closely parallels that of the dog which kills at about three-day
intervals.

Mitchell (1916) also examined the gut and confirmed
Crisp's findings, noting that the gut was the same internal
diameter throughout.

Reproductive system
The unusual aspects of reproduction in marsupials attracted
much attention in Europe during the early days of the exploration of Australia, but it was not until 1953 that Pearson
& de Bavay described the female reproductive system of the

thylacine in some detail, although Beddard (1903) had described the structures to be seen in transverse sections of the ovary. The fine details of these structures need not concern us beyond noting that the arrangement of the uterus, the uterine neck, and the medial vaginal cul-de-sac bears a close resemblance to that seen in the Tasmanian devil. There are some minor differences in the structures such as a very long urinogential sinus in *Thylacinus* and a shorter uterine neck in *Sarcophilus*. However the resemblances in this system suggest a familial relationship between the thylacine and the Tasmanian devil and is further evidence to support the view that the thylacine's closest relation is the Tasmanian devil as the arrangement of the reproductive system does not closely relate to that of the smaller marsupials such as *Antechinus*, the marsupial mouse.

The little that we know of reproduction in the thylacine will be described in Chapter 5.

There may be small quantities of preserved soft parts in small collections throughout the world. R.L. Hughes (pers. comm.) recently discovered a preserved thylacine reproductive tract in a collection in Holland, and two female genital systems are preserved at the Institute of Anatomy, Canberra, together with some thirty-five other parts from at least three individual thylacines.

Pouch young

The young are carried in the pouch of the female for the first few months of their life. In the first couple of weeks the young are difficult to distinguish from those of any other marsupial but with the development of vibrissae they can be identified. Further details of the externals of the pouch young have been described by Boardman (1945); and more information is provided in Chapter 5.

Comparison with other marsupial and eutherian predators

The thylacine occupies a unique position amongst the living marsupials as it is the largest of the predators. Other marsupial carnivores are medium-sized like the Tasmanian devil, or small like the tiger cat and native cat, and some are tiny such as the marsupial mice.

Moeller (1968) gives a very detailed comparison of the anatomy of *Thylacinus* with that of other Dasyure marsupials. He shows that the body proportions are similar to those of *Dasyurus* but are different from those of the tiger cat, *Dasyurops*, and from those of *Sarcophilus* in that the rear limbs are longer. The skull length plotted on a logarithmic scale against the length of the fore and rear limbs shows that these proportions in *Thylacinus* resemble those of the native cat and the tiger cat, falling on the same straight-line relationship, whereas in the Tasmanian devil the slope of the curve is different.

These conclusions are not only of academic interest but reflect the life form of each species. The native cat is an active predator whilst the tiger cat, although not as active, nevertheless leads a predatory existence which requires a considerable degree of agility. The Tasmanian devil on the other hand is a slow-moving scavenger requiring little agility and this is shown in the relatively poor development of the rear limbs. The thylacine being an active predator requires considerable power in its back legs and this is shown in the limb proportions.

Moeller then goes on to compare the anatomy and body proportions of the thylacine with those of some selected cursorial canids, namely the wolf, *Canis*, the Indian wild dog or dhole, *Cuon*, the striped hyaena, *Hyaena*, and the maned wolf, *Chrysocyon*. He found that the forelimb length of *Thylacinus* is shorter than that of the canids except *Cuon* which has a shorter forelimb than other canids and so resembles *Thylacinus*. The neck of both the thylacine and the dhole is short.

The long tail of the thylacine is much longer than that found in any of the canids and is rigid. The long tail is a feature commonly found in marsupials, very few of which have a reduced tail or no tail.

Moeller points out that the cursorial felids, particularly the serval and the cheetah, have body proportions similar to those of the canids, and then concludes that the body proportions of the thylacine more closely resemble those of the leopard than those of the canids. Both the thylacine and the leopard have short legs and long tails, but the leopard has a shorter skull. This intriguing morphological similarity does not mean that the thylacine has the same habits as the leopard, the similarity being quite fortuitous. The notion of thylacines climbing trees with large bits of prey is quite attractive but is not supported by even the most imaginative 'tiger tale'.

Moeller's study shows that the thylacine is not as fully adapted to the cursorial life as the canids, and his main conclusion is that there is no real relationship between the body proportions of the wolf and those of the thylacine. Any resemblance is superficial although there is some similarity in the shape of the foot pads.

Moeller then contrasts the skull of *Canis* with that of *Thylacinus* and shows that, apart from the basic differences between marsupial and eutherian skulls, the nasal bones and the squamosal arches are different and the male *Thylacinus* has a much wider forehead and jugal width than the female. We have already seen that Krefft erected his new species upon these skull differences. The forehead width of the thylacine is greater than that of the wolf but there is very little difference in jugal widths. The length of the cranium to the midpoint of the skull is greater in large wolves than it is in large thylacines. The major difference in the morphometrics of the skull of the two species is that the width at the penultimate molar of wolf skulls is greater than in those of thylacines by 8 mm, or more in large animals.

The brain case is about 14 mm longer in wolves and the capacity of the braincase in wolves, 100 mL, is very much greater than in thylacines, which only contains 60 mL. The brain volume/length of animal curves show different slopes indicating little growth of the brain in older thylacines.

The dental formula of the thylacine is $\frac{4}{3}$; $\frac{1}{1}$; $\frac{3}{4}$; $\frac{4}{4}$ and that of the wolf $\frac{3}{3}$; $\frac{1}{1}$; $\frac{4}{4}$; $\frac{2}{3}$. Moeller notes that the carnassial teeth of the thylacine are weaker than those of the wolf and this may be associated with the smaller prey which the thylacine kills.

The comparison with the wolf was carried further by Keast (1982) who compared the body proportions of the thylacine with those not only of the wolf but also of the tiger cat, marsupial mouse, weasel, and marten. He found amongst other things that the metatarsal and metacarpal segments are proportionally shorter in relation to the length of the limb in *Thylacinus* than in the placentals. He goes on to conclude that although there is a superficial resemblance in the appearance of the wolf and the thylacine, the only body proportion in which the thylacine resembles the wolf is the increased length of the neck relative to other Dasyures. This will be mentioned further in relation to locomotion (Chapter 5). Keast also concluded that although thylacines were capable predators, they were not as specialized as the wolf in this role.

In conclusion, the adoption of the cursorial hunting habit has led to close superficial resemblances between the wolf and the thylacine but upon analysis the body proportions are quite different. The closest resemblance is in the skulls but the brain of the wolf is very much larger. The general build and body form of the wolf give it an agile appearance while the thylacine appears slower and clumsy with an ungainly heavy tail.

4

Thylacines in zoos

The rarity of thylacines and their extraordinary superficial resemblance to wolves led to some demand for these animals for exhibition in zoos. However, although local dealers would have been keen to supply thylacines for overseas zoos, not many animals were exported. Only three Australian zoos appear to have acquired thylacines and six overseas collections had specimens on show at one time or another. The advent of steamships and the opening of the Suez Canal made the transportation of live animals to Europe more rapid and animals could be shipped with greater confidence than previously.

The earliest display of thylacines in a zoo was in 1843 when the then Governor of Tasmania, Sir John Eardley-Wilmot, maintained three animals in a private collection in the Government Gardens, now Franklin Square. Sir John was recalled in 1846 and we hear no more of the collection after that.

Table 4.1 shows that the largest exhibitor of thylacines either overseas or in Australia was the London Zoo, which readily purchased the animals, and was the first to display them. The London Zoo exhibited seventeen thylacines in all

TABLE 4.1 Thylacines displayed at zoos

Zoo	Sex	Date of		Longevity	Source	Authority
		purchase	death			
Adelaide	–	1886	1890	4y		National Parks file[a]
	–	1889	–	–		National Parks file[a]
	–	1891	–	–		National Parks file[a]
	–	1897	–	–		National Parks file[a]
	–	1898	1903	5y		National Parks file[a]
Antwerp	–	–	–	–		National Parks file[b]
Beaumaris (Hobart)	–	18.6.1910	–	–	Blackwood	Roberts diary[c]
	F	2.5.1912	–	–	Tyenna	Roberts diary[d]
	M	2.5.1912	–	–	Tyenna	Roberts diary[d]
	M	10.1912	9.3.1915	2y 5m	Tyenna	Roberts diary
	M	23.10.1915	–	–	Harrison	Roberts diary
	–	9.6.1916	–	–	Harrison	Roberts diary
	M	17.6.1916	–	–	Tyenna	Roberts diary[e]
	M	3.6.1917	–	–	Harrison	Roberts diary
	M	30.6.1917	–	–	Harrison	Roberts diary
	M	3.7.1923	–	–	Slater	HCC 1923[f]
	F	19.2.1924	–	–	Mullins	HCC 1924[g]
	–	19.2.1924	14.4.1930	6y 2m	Mullins	HCC 1930
	M	19.2.1924	1935	11y	Mullins	HCC 1935
	F	19.2.1924	7.9.1936	12y 7m	Mullins	HCC 1936
	F	21.7.1925	–	–	Murray	HCC 1925[h]
Berlin	M	1902	1908	6y	–	Collins 1973
Cologne	–	–	–	–	–	National Parks file[b]

(continued)

Zoo	Sex	Date of purchase	Date of death	Longevity	Source	Authority
London	M	9.4.1856	–	–	–	Matthews 1958
	M	2.5.1863	–	–	R. Gunn	Matthews 1958
	F	2.5.1863	–	–	R. Gunn	Matthews 1958
	M	14.11.1884	2.4.1893	8y 5m	dealer	S.S. Flower 1931[i]
	F	14.11.1884	5.2.1891	6y 3m	dealer	Matthews 1958[i]
	M	19.3.1886	–	–	deposited	Matthews 1958
	F	19.3.1886	–	–	deposited	Matthews 1958
	M	30.6.1888	–	–	deposited	Matthews 1958
	F	30.6.1888	–	–	deposited	Matthews 1958
	M	24.4.1891	–	–	exchange	Matthews 1958
	F	24.4.1891	–	–	exchange	Matthews 1958
	M	26.3.1902	17.1.1906	3y 10m	purchase	Matthews 1958
	F	12.3.1909	5.12.1914	5y 3m	Beaumaris	Matthews 1958
	M	18.4.1910	20.11.1914	4y 7m	Beaumaris	Matthews 1958
	M	21.11.1910	25.12.1914	4y 1m	Beaumaris	Matthews 1958
	M	21.11.1910	20.11.1912	2y	Beaumaris	Matthews 1958[i]
	F	26.1.1926	9.8.1931	5y 7m	purchase	Matthews 1958
Melbourne	–	1875	–	–	–	National Parks file[k]
	–	1888	–	–	–	National Parks file[l]
	–	1894	–	–	–	National Parks file[m]
	F	1899	–	–	–	National Parks file[n]
	–	1906	–	–	–	National Parks file[o]
	–	1925	–	–	–	National Parks file[p]
New York	–	17.12.1902	15.8.1908	5y 8m	–	Doherty (pers. comm.)
	–	26.1.1912	20.11.1912	10m	–	Doherty (pers. comm.)
	–	7.11.1916	13.11.1916	6 days	–	S.J. Smith 1980
	–	14.7.1917	13.9.1919	2y 2m	–	S.J. Smith 1980

Smithsonian (Washington)	–	7.5.1904	10.4.1909	5y 3m	–	Collins 1973
	M	3.9.1902	18.9.1902	9 days	–	S.J. Smith 1980[q]
	F	3.9.1902	13.10.1909	7y 1m	–	Collins 1973[q]
	M	3.9.1902	10.1.1905	2y 4m	–	Collins 1973[q]
Taronga Park (Sydney)	–		12.10.1918	–	Beaumaris Zoo	Roberts' accounts[r]

[a] Adelaide obtained a pair in each of these years and four were on display in 1890. Their disappearance from the collection apparently was a cause of little regret as the animals were not of much public interest.

[b] Nothing further is known about either of these animals.

[c] Blackwood lived in Fingal and caught the animal near there.

[d] These were described as 'a pair'.

[e] Sold, probably to Ellis Joseph of Sydney for £25 on 14.7.1917 (Roberts diary).

[f] Cost £25.

[g] The female had three cubs and the lot cost £55.

[h] Cost £25.

[i] These were purchased from Crowther of Launceston (Renshaw 1938).

[j] Sold to New York Zoo on 10.11.1912 for £80 (Matthews 1958).

[k] Three animals were obtained. No further details available from the scanty records for this zoo. There were at least twelve thylacines on display at one time or another but probably more were shown.

[l] Five animals. No details.

[m] Thylacines were on exhibit between 1894 and 1906 but the only dated accession was in 1899.

[n] Female with four cubs.

[o] Not clear whether some new stock was obtained or whether the display ceased.

[p] Unknown number obtained. Died 1931. Was this the Churchill specimen (see p. 59)?

[q] These three were raised from a litter.

[r] The records of Taronga Park do not make note of this thylacine (Bell 1965) but it is clear from the Roberts' diary: 12 October 1918 'shipped tiger to Sydney Zoo', and 19 April 1919 received £25 for 'Taronga Park tiger'.

and had one or more on display from 1856, probably almost continually until about 1868. Then there was a gap until 1884 when more thylacines became available and they were on display until 1906, and again from 1909 until 1919, and their last animals were on show 1926–31. In 1953 at the time of the coronation of Queen Elizabeth II, I remember the Fauna Board receiving a request from the Zoological Society of London for some marsupials to form a 'coronation collection'. The request list included a thylacine, so even as late as this the London Zoo was still optimistic—perhaps more than hopeful—of having a thylacine on display again.

The Beaumaris Zoo at Hobart was the second largest exhibitor, having had twelve thylacines on display at various times after 1910. Beaumaris Zoo made a feature of this display from its commencement in 1895 and it is reasonable to estimate that between 1895 and 1910 at least five thylacines would have been exhibited. Other thylacines passed through this zoo for trading purposes and these are listed in Table 4.2. There would have been animals on display prior to 1910 but the figures are not available as the zoo records before 1910 have been lost. The lists given in the tables are reasonably complete and contain records of the Beaumaris Zoo which have not been collated previously. The Beaumaris records are important as they provide another source of information about the longevity of the species.

Thylacines in captivity did not provide much entertainment as they apparently did little to please visitors, looked stupid, and seemed very bored with the whole proceedings (Harris 1808; Renshaw 1938; Grzimek 1976) but in spite of this it is clear from advertisements that some zoos were keen to have specimens and attempts were made to supply them (*Hobart Mercury* 1874).

The animal which died in the London Zoo in 1931 was the last thylacine known to be on display in an overseas zoo. Its Tasmanian origins are unknown but the Zoo purchased it from an animal dealer, G.B. Chapman of Tottenham Court Road, for £150 (Matthews 1958). The last thylacine to die in captivity was that in the Beaumaris Zoo which died on 7 September 1936 (HCC 1936). The carcase was sent to the Tasmanian Museum but the skin was in such poor condition that it was useless. For some reason which is not clear to me it has been generally accepted that 1934 was the date of the death of the last captive thylacine but the records of the Hobart City Council are quite positive on this point.

The two principal exporting agencies were the Beaumaris Zoo and J. Harrison of Wynyard (northern Tasmania), an animal dealer. There may have been others but I have found no trace of their dealings in my searches. For example, Doherty (1977) describes the capture of a thylacine about 1912 which was sent to Prof. T.T. Flynn at the University of Tasmania who sold it (for £40?) to Sydney. It is not known to whom he sold it or where it went, but it did not go to Taronga Park Zoo. It may be the animal which turned up in the New York Zoo on 26 January 1912, as the time factor is about right since Flynn was at the University of Tasmania at this time. If it had been exported through Sydney, then Ellis Joseph would have been the most likely exporter.

About 1925 Elias Churchill caught a female and three cubs in the Florentine Valley. These were not sold but were carted around Tasmania and displayed for a charge of sixpence per person (Brown 1983).

The three specimens obtained by the Smithsonian Institute, Washington, in 1902 are of more than passing interest as both Collins (1973) and S.J. Smith (1980) refer to them as 'arriving as pouch young'. This is fascinating, and what happened to the mother? Surely she must have arrived at the same time but there is no record of her arrival at the Smithsonian Zoo. How were the young reared and how old were they when they first arrived in Washington? One of the young died after only nine days but the others were successfully reared, one of them surviving for more than seven years. This is the only known record of the successful rearing of pouch young in captivity and the only record of the rearing of young without the benefit of the mother's milk and pouch.

The animal sent to London Zoo by Gunn in 1863 had three young in the pouch when caught but there is no record of them arriving in London. It is likely that they were removed before shipment or they may have died en route.

Beaumaris Zoo obtained a female with three cubs from

Mullins on 19 February 1924. The term *cub* is used to describe the young at foot and is not applicable to pouch young. These cubs were raised and one of them was the last thylacine to died in captivity in 1936, although Brown (1983) suggests that this last animal came from Churchill of Ouse but the Roberts' diaries make no mention of this.

Longevity

An animal purchased by London Zoo in 1884 lived for eight years five months and this is the longest known time for a thylacine in captivity in an overseas zoo. It would have been perhaps a year old when purchased from a dealer and this suggests a life span of nine years six months.

The thylacine obtained by Beaumaris Zoo from Mullins in 1924 did not live very long but one of her cubs survived until 1936, another dying in 1935. This provides the longest known life span of any thylacine, some twelve years.

Many animals live longer in captivity than in the wild but the thylacines's relative the Tasmanian devil lives longer when in its natural state than in captivity. S.S. Flower (1931) reported a longevity of five years nine months for this species but we know from field trapping that many individuals live longer than that, reaching up to eight years (Guiler 1978). If the thylacine follows the same pattern, a longevity of 12–14 years can be anticipated in the wild.

Thylacines in Beaumaris Zoo

The Beaumaris Zoo in Hobart was started by Mrs Mary Roberts in 1895 in the extensive grounds of her home on Sandy Bay Road, and she ran it until her death in 1921. Her daughter gave the Zoo to the Hobart City Council which placed it under the Reserves Committee and appointed Mr A.R. Reid as Curator. The collection was moved within two months to the Hobart Domain where it continued to operate until 31 October 1937, when it was closed because the City Council would not sustain the financial loss any longer.

Beaumaris Zoo was well placed to develop a thylacine export trade as trapped animals could be settled into captivity before shipment. The animals could be sold or exchanged on a zoo–to–zoo basis or retained for display.

TABLE 4.2 Thylacines purchased and exported by Beaumaris Zoo, 1910–37

Date	No.	Sex	Locality/Source	Cost	Fate
Purchases					
18.6.1910	1		Fingal, Blackwood Bryant	£8	Dead on arrival.
6.6.1911	1			–	Both died, 1913.
2.5.1912	2		Tyenna, O'May	£20	Both died 13.5.1912
6.5.1912	2	F, M	Launceston	£20	
10.1912	1		Tyenna, O'May	–	
23.10.1915	1		Harrison	£17–10	
9.6.1916	1		Harrison	£18	
17.6.1916	1		Tyenna, O'May	£12	
3.6.1917	1	M	Harrison	£20	
30.6.1917	1		Harrison	£20–1	Returned as it had an inflamed foot.
3.7.1923	1		Slater	£25	
19.2.1924	1	F	Mullins	£55	Had 3 cubs, see Table 4.1.
21.7.1925	1		Mullins	£25	
Exports					
2.3.1910	1		London Zoo	£stg40	Ex pre-1910 stock. Lived 4 years in London.
3.10.1910	2		London Zoo	£stg68	See Table 4.1.
10.1911	1		London Zoo	–	Not included in London Zoo displays (Matthews 1958). The Roberts diary records the export per s.s. Persic.? to New York, 26.1.1912.
24.1.1912	1		London	£stg30	Also not in London Zoo lists.
14.4.1917	1		Joseph, Sydney	£25	? to New York 14.7.1917
12.10.1918	1		Taronga Park	£25	Not recorded by Taronga Park. See text.

Note: Stock on hand on 1 January 1910 estimated to be 4 at least.

In compiling this section I have been given access to Mrs Roberts' surviving diaries and cash books very generously by her grandson, Mr Gerald Roberts. These records cover the period 1910–21; the earlier ones are missing. The Hobart City Council kindly allowed me access to the minutes of the Reserves Committee and of the Council from 1921 until the closure of the Zoo in 1937.

Mrs Roberts' Records do not extend to the history of each animal, but the purchase or sale of a thylacine was of sufficient importance for an entry to be made in the cash book or diary, or both. The City Council records are more detailed and it is possible to trace the history of individual animals. However, it should be remembered that Harrison of Wynyard also exported thylacines during all of this period but I have no records of his dealings.

From 1910 to 1937 the Beaumaris Zoo received sixteen thylacines, seven of which were exported to Australia or overseas (Table 4.2), one was dead on arrival, and another was returned to its captor as it had an inflamed foot. Probably more of these were traded than were ever on display in the Zoo. Between 1910 and 1921 nine animals were recorded as having been purchased and seven of these exported, five in 1910 and 1912, and two in 1917 and 1918. At one time there were no thylacines on exhibition, a diary entry of 9 March 1915 recording 'the only tiger found dead', and it was not until 23 October 1915 that one was bought from Harrison of Wynyard. It is clear that by this time Beaumaris Zoo could no longer readily obtain thylacines for export from the usual channels as it had done earlier.

Some of the thylacines at Beaumaris may have been used for exchange as Grzimek (1976) and Sharland (1966) both make this claim, the latter author recalling that a polar bear, a pair of lions, an elephant, and other animals were obtained by this means.

In addition to dealing in live thylacines, the Zoo sold carcases and skeletons of those that died. Mrs Roberts offered a carcase to the Australian Museum on 10 March 1915—this was the male animal from Tyenna which had been caught in October 1912. A Mr Hardy offered £10 for a skeleton (19 May 1915) and Professor Flynn obtained skeletons (1 October 1918).

Thylacine, Beaumaris Zoo, Hobart. *Source*: State Library of Tasmania.

Mrs Roberts made constant attempts to obtain thylacines captured and held by other persons. Her diary entry of 4 August 1910 records 'sent telegram to W. McDonald, Irishtown, about half-grown tiger' and Ellis of Fingal was approached with a similar request on 10 March 1915, but nothing came of these efforts.

We can assume that other persons besides Harrison and Mrs Roberts were catching, displaying, and probably exporting thylacines by this time as they had become scarce. A case in point is the fate of the three cubs caught at Britton's Swamp with their mother by Mr Rowe sometime in 1910–12. The female died but the cubs were kept for four months and then sold to J. Harrison for £5 each (Bell 1965), one going to Launceston and two to Hobart or Melbourne. Further confusion is seen in the case of a family of thylacines caught in north-west Tasmania, Mrs Roberts' diary on 14 March 1910 records sending £3 'as an extra payment for the man who caught the family of tigers'. The money was not sent to either Rowe or Harrison but to Mr McGaw who was the manager of the Van Diemen's Land Company at Woolnorth, so it would appear that these particular thylacines came from there and were not the Rowe cubs. On the other hand, Mrs Roberts' accession records during this time show only the purchase of three thylacines from O'May of Tyenna, and there is no evidence that she ever received the thylacines from Woolnorth.

The person who received the £3 may have been G. Wainwright, the last 'tiger man' on Woolnorth. He recalled when interviewed by me in 1961 that he caught three thylacines at Mt McGaw and sent them to Hobart Zoo. Wainwright remembered the date as 1913 but this may be incorrect. There are no records of any of these animals in the Roberts' documents.

By about 1912 the catching of a thylacine or cubs was an event worthy of note, but there are many discrepancies in the few existing records and especially about what happened to the animals. They were fetching a good price and various persons were buying them and either exporting them to unknown destinations or exhibiting them around Tasmania.

The intermediaries made money out of their sales. The Beaumaris thylacines were bought for about £A20 and sold to London for £stg40, a profit of $A28. The London Zoo purchase in 1926 from Chapman at £stg150 must have given him a very good profit, wherever he obtained his animal, as the Tasmanian price at this time was about £A40–50.

Beaumaris Zoo under the Hobart City Council

When the Hobart City Council took over the Beaumaris Zoo in 1921 there was only one thylacine on the accession list (HCC 1922) and this situation lasted until 1923 when the Slater animal was bought, followed in subsequent years by the adult and litter of three bought from Mullins, and the Murray female bought in 1925. There were no purchases after this date.

Three deaths occurred in the 1930s. The first of these animals died on 14 April 1930; it was six years old and succumbed to 'kidney disease' (HCC 1930). The carcase was sold to the Tasmanian Museum for £5. This animal was probably one of the Mullins cubs. The second, a male, died on 3 July 1935, and it was one of the Mullins cubs. It was attended by a veterinarian who diagnosed pneumonia, and its skin was in such poor condition as to be useless. The third died on 7 September 1936 (HCC 1936) and the body was sent to the Tasmanian Museum. This also was a Mullins cub and was the last thylacine to be on exhibition. The HCC records show a discrepancy between the number of animals purchased and the number dying. There was only one in stock in 1921 and three adults plus three cubs were purchased but only five deaths are recorded, the other animal must have died between 1932 and 1934, a period for which the records are missing.

Attempts were made by the Zoo to replenish the thylacine collection, the sum of £15 being offered in 1923, but £25 had to be paid to get an animal. The offer was raised to £30 in 1936 and a final offer of £40 was made in 1937 (HCC 1937) but this did not produce an animal. A Mr Chaplin offered to catch a thylacine for the Zoo provided that he was paid £4 per week for two men for a month together with £2 for food, but the offer was not accepted.

The scarcity of thylacines after 1921 effectively prevented the Zoo from using them for exchange or trade, and although there was a standing offer, no animals were forthcoming. The Fauna Board refused to issue a collecting permit to the Zoo in the attempt to obtain a thylacine for £40 in 1937, and the Board was by this time on the point of launching the first search to determine if thylacines still existed.

As early as 1928 an attempt was made by some members of the Fauna Board to obtain protection for the thylacine, not solely on account of its rarity but because 'prices paid by Mainland and other institutions make it impossible for Tasmanian Museums and Zoos to obtain specimens' (Animals and Birds Protection Board files H/60/34). Applications for permission to export thylacines were considered by the Commonwealth Advisory Committee until 29 August 1933, when the power was transferred to the Fauna Board. The Board was concerned about the future of the species and a subcommittee was established on 19 September 1933 to consider the matter. In spite of this concern, on 21 August 1934 a permit was granted for a pair of thylacines to be captured for breeding in Melbourne Zoo, and another was granted on 10 December 1935 to Prof. Berkitt of Sydney University to catch a thylacine. Although there is no minute of this, the

Board must have decided against the overseas export of thylacines as an application from the Belfast (Northern Ireland) Zoo for a thylacine to be obtained from the Beaumaris Zoo was refused on 12 February 1935 (HCC 1935).

The only other animal collection in Tasmania was that in City Park, Launceston. There were thylacines on exhibition at times but the number is not known (King in thylacine file, State Archives of Tasmania). The fleas described by Dunnet & Marsden (1974) probably came from a thylacine on display in Launceston in 1879. Thylacines were exhibited at the Launceston Show about 1910, probably from the City Park collection.

There were some thylacines kept on local temporary display. For example, a fine male specimen was to be seen at George Warsden's Livery Stables in Hobart in the 1880s and a female with three cubs was on display in the London Tavern in Launceston in 1864 (*Hobart Mercury* 1864). A Mr Bart of Tenalga kept two in a shed. He bought them from snarers and eventually sold them to City Park, Launceston about 1912. A Mr Dodery caught four in a pit trap, the two old animals he killed for bounty and the other two were sold to Sydney.

It would be interesting to know how many thylacines died before reaching these zoos. Early trappers spoke of them dying 'shortly after capture or being prone to shock' (Hayes 1972). Hayes goes on tell that G. Stevenson said that they died readily in a snare and records that one animal which had been shot in the tail died of shock shortly afterwards. Both G. Smith (1909) and Sharland (1957) say that thylacines 'gave up' when caught in a snare. I would suggest that those animals listed in the tables in this chapter are but the survivors of a much larger band who failed to survive the initial trauma of capture.

There was an extraordinary apathy shown by the various zoos about the fate of the thylacine. There is no record of any attempts being made to breed them in captivity, nor do we find any expressions of concern for their future.

5

Some facts and some deductions

Pathetically little is known of the biology of the thylacine. It is possible to make reasonable deductions from the early anatomical descriptions but we have to complete the picture from the other sketchy information available. However, this is not made easy for us as the literature contains some contradictory statements and some observations are obviously incorrect.

We must not condemn our predecessors for not investigating a species that was both interesting and, even then, rare enough to excite curiosity: we must consider this apparent failure in the context of the times. The only local research institutions in existence when thylacines were abundant were the Tasmanian Museum and the Queen Victoria Museum in Launceston. Both of these museums had very small staffs and employed no trained zoologists. Both museums, as was typical of all such places in the post-Darwinian period, were concerned with amassing collections of animals which were preserved and classified before being put on display to the public. The Foundation Chair of Zoology at the University of Tasmania was not established (at that time as a lectureship) until 1909, by which time the thylacine decline had taken place and specimens were difficult to obtain.

At the beginning of this century overseas research institutions advertised for and requested thylacines, but their demands could seldom be met as the animals were already scarce, and trappers could probably collect more money from a carcase by taking it around various properties than by offering it to scientific institutions.

No one ever contemplated field research programmes on the species, and the thylacine had almost disappeared by the time the science of ecology was being developed in Tasmania. Even as late as the 1950s I well recall the astonishment with which trappers observed our collections of possum material, especially when it came to the removal of the testes.

Nevertheless there are records of a small trickle of thylacines, even as late as the 1920–30 period, coming into the museums and zoos and the University of Tasmania, and some justifiable criticism can be directed at these and other research institutions for not making better use of this material for research, when it was obvious that there had been a drastic reduction in the thylacine population, and it was known that the animal no longer existed on the Australian mainland.

The role of zoos in the community has changed over the years, and the modern zoo is concerned not only with displaying animals to the public but with the conservation of species and re-establishing them in their former habitats. It is a pity that this change came too late to rescue the thylacine.

Former distribution in Tasmania

The early diarists recorded only brief details of the localities in which thylacines were seen. For example, the Robinson diaries (Plomley 1966) record that thylacines were found on the coast between the Mersey River and Port Sorell and they were frequently encountered in the vicinity of Hampshire. Robinson's party saw three thylacines on the coast between Sandy Cape and Cape Grim. The various Van Diemen's Land Company records give us useful information on the microdistribution of thylacines (Chapter 6) and are one of our most reliable sources.

Good thylacine country: Mount Projection, Western Tiers.

To study the biology of a species, it is necessary to have a large number of samples and this involves killing the animals. It is ironic that the government bounty scheme provided a source but practically no use was made of the material at the time, although the details of the bounty payments provide distributional data which otherwise would never have been gathered.

A total of 2063 thylacines upon which a government, Van Diemen's Land Company, or Hamilton Municipality bounty was paid could be traced to their locality of capture. Some latitude and discretion has to be taken in the interpretation of the data. For example, there were sixty claims made by residents of Ross and the surrounding district but it is highly unlikely that these thylacines were caught in the immediate vicinity of Ross. Talks with some old residents of the town many years ago revealed that the animals were mostly caught in the hills to the east of the town or in the foothills of the Western Tiers.

There were few thylacines caught in the vast tract of

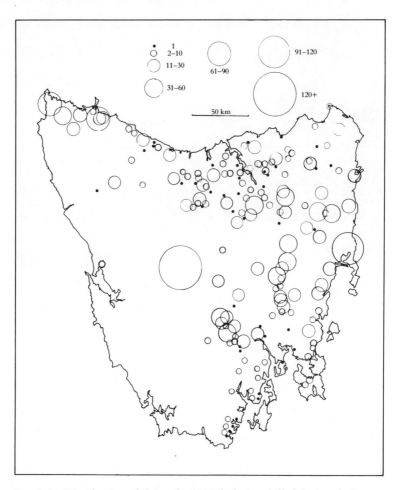

Fig. 5.1   Distribution of claims for 2131 thylacines killed during the bounty period, 1888–1909 (modified from Guiler 1961a).

mountainous country from Macquarie Harbour to Lake St Clair–Cradle Mountain–Waratah. The thick rainforests covering much of this country did not support large game populations, the terrain is rugged and mountainous and was not suitable for sheep grazing, and is much the same today. A few thylacines may have been caught on the buttongrass plains between the mountain ranges in this desolate country.

Fig. 5.1 shows that thylacines were abundant throughout

Tasmania with the exception of the west coast ranges and the south west. The central plateau produced the largest number of thylacines for bounty, over 660 claims being made from this area. The region was, and still is, rich in game and was the most important scene of winter snaring for the fur industry. It was extensively used for sheep grazing, especially in summer. Within the central plateau, the greatest yield of thylacines came from the Dee Bridge–Derwent Bridge district, which produced 233 animals, mostly caught by the Pearce, Jenkins, and Stannard families. The fringes of the plateau, particularly on the north-eastern boundaries near Cluan, Mole Creek, and Deloraine, were highly productive.

Another large population lived in the Mt Barrow–Ben Lomond highlands and the surrounding woodlands. This large concentration of thyacines extended south-eastwards towards the east coast and also in a north-easterly direction towards Gladstone.

Apart from the ninety-one thylacines taken in the Bicheno–Cranbrook area there were relatively small numbers produced for bounty payments in the regions on the east coast where it was alleged that high thylacine predation took place, Swansea having only seventeen claims, Lisdillon twenty-two, and Little Swanport twenty-three. All of these animals would certainly have been caught in the hill runs in the ranges between the coast and the Midlands Highway.

Large numbers of thylacines were caught in the north west, most coming from Woolnorth and Smithton. The Smithton catch would contain all those thylacines caught in country districts and collected by the local carter (see Chapter 6), and a substantial number of them probably came from Woolnorth or adjoining properties as well as from further down the west coast as far as Sandy Cape.

South west Tasmania in the bounty times was largely unpopulated except for men working there prospecting, felling timber, and track cutting. In spite of the financial incentive not many thylacines were taken in the south west, the late Mr H. Reynolds, who packed supplies along the Port Davey track, told me that thylacines were not common and were found only on the buttongrass plains and on the coast. He said that any thylacines that were caught there would probably have been taken to Tyenna for bounty payments, then the entry point for the south west.

Both the Tasman and Forestier Peninsulas formerly supported thylacine populations, bounty claims being lodged at Nubeena, Bream Creek, and Dunalley. Mr T. Dunbabin of Bangor spoke of the Ragged Tier as being particularly favoured.

The distribution of thylacines bore no relation to altitude. They were found throughout the state, and, if anything, favoured the coastal plains and scrub. However, open savannah woodland was used extensively by thylacines and they were not 'confined to the mountainous regions' as was so frequently stated in the literature. This was substantiated by all of the old trappers I talked to about thylacines. For example, H. Pearce told me that all the thylacines caught by his family were taken in the country lying to the east of the King William Saddle and they never caught any on the other side of the King William Range. These mountains form the beginning of the west coast ranges with their associated rainforests, whereas the country trapped by the Pearces is open woodland with plains and abundant game.

Catching a thylacine was quite an event even when they were plentiful, and in talking to old-timers I would be told at times every detail down to the very fence where he set his snares to catch the animal.

Reproduction

The first step taken for the protection of the species was in 1929 when a closed season was declared for December, which was believed to be the month when mating took place. Neither Asdell (1946) nor Zuckerman (1953) give any details of the breeding season, and as far as I am aware, mating of thylacines has not been observed either in captivity or in the wild.

By 1860 the statement had come into the literature that thylacines retired to the mountainous and rough inaccessible places to breed (Gould 1863), but by this time they were confined to that type of country and were uncommon in the settled parts of the state. Later in the century when thylacines were plentiful in all pastoral areas there is no evidence of this migration. Neither of the other two Dasyures, the native cat and the Tasmanian devil, move from their usual home range for breeding.

Lord (1928) states that thylacines laid a scent trail across country during the breeding period. However, although it is not impossible that some sort of 'marking' may occur, the literature does not contain descriptions of well-developed rectal or other glands suitable for this activity.

Several authors have claimed that thylacines have a lair or other retreat and that the young are brought up there, the lair being variously described as a hollow log, hollow tree, a cave, or a rock cavity (Lucas & Le Souef 1909; G. Smith 1909; Le Souef & Burrel 1927). There is no doubt that a thylacine will have its favourite sleeping places but this is very different from establishing a den in which the young are brought up.

Pouch young
The number of young carried in the pouch varies between one and four, the latter being the number of nipples available. The few sources do not enable us to derive a mean number of pouch young per female, but we can deduce that the thylacine is polyovular and in this follows the usual Dasyure pattern. The number of eggs produced by the ovary during one season is not known. Observations on other Dasyures, notably the native cat, *Dasyurus quoll* (Hill 1900), and the Tasmanian devil (Guiler 1970; Hughes pers. comm.), have shown that many more ova are produced into the uterus than are required to fill the pouch seating space. Hughes has found up to 100 fertilized ova in the uterus of a Tasmanian devil but the fate of these has not yet been determined; it is interesting to speculate whether all reach full term or are resorbed at some time before complete development. If all reach full term and are born, then there must be a very high mortality during the struggle for the four teats.

It is unlikely that a female in her first breeding year would have four young, two is much more likely, and three or four would be reached in the peak breeding years, followed by a decline as the animal grows old.

The gestation period is not known but if this follows the

usual Dasyurid pattern it should be about thirty-five days. The length of time that the young spend in the pouch was given by Le Souef & Burrel (1927) as three months (about ninety days). This seems to be very short compared with many other species of marsupials, in which the pouch young do not reach a stage when they are fully furred and their systems become fully functional and they cease to be an advanced embryo and become a baby mammal until at between 100 and 110 days. However, they are not yet capable of assuming an independent existence. By about 140 days a young Tasmanian devil is very mobile and capable of running about but is still being nursed by the mother (Guiler 1970). This is close to the four and a half months as quoted by Roberts (1915) for the Tasmanian devil. If we assume that the breeding biology of *Thylacinus* is similar to that of the Tasmanian devil then a pouch life of 130–140 days could be expected, followed by a period of a month or so when the young are hidden by the mother while she hunts. During this time the litter would still be largely dependent on milk but in the next phase when they run after the mother the process of weaning would commence.

A pouch young of 7.5 mm crown–rump length was described by Boardman (1945) as being naked, devoid of colour pattern, and showing no development of the fur. A larger young of 288 mm was furred and its eyes were open. It would have been about full term. The size of the full-term pouch young is greater than in other Dasyures, the Tasmanian devil young being about half this size.

It is possible to glean some information on breeding from the Lands Department account books. The bounty for young animals, but not pouch young, was ten shillings and a total of 150 subadult animals was submitted for bounty (Fig. 5.2). The clerks differentiated between pups and half-grown young in about two-thirds of the claims.

Most of the pups were found in the winter months while most of the half-grown young were submitted about one month later. All three of the curves in Fig. 5.2 show a bimodal distribution which probably is an artefact due to the small sample. It is intriguing that all three curves show this distribution.

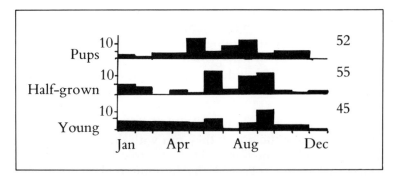

Fig. 5.2 Monthly distribution of the catch of 152 young thylacines. The young are graded in two sizes, namely pups and half-grown, while the third category refers to undefined young (see text). Modified from Guiler 1961a.

Some breeding appears to take place throughout the year, both pups and half-grown young being recorded in every month, and this is substantiated by Pearce's remarks that young could be found in the pouch at all times of the year.

The Dasyures all tend to have a fairly sharply restricted breeding season, for example the Tasmanian devil mates in March–April (Guiler 1970). The breeding season is tuned to the assumption of a free life by the young at a time which is most advantageous to them, namely late spring or early summer whenever food and mild weather give the young the best chance to become established. The reproductive organs regress during the non-breeding season so that it becomes impossible for reproduction to take place. Nevertheless some out-of-phase breeding has been recorded as taking place in the Tasmanian devil, both Green (1967) and Guiler (1970) finding some pouch young which were up to five months out of phase with the breeding in the community.

It is assumed from all accounts that thylacines breed once a year and this is in line with the general pattern of reproduction in the Dasyures. The species is monoestrus and there is no evidence of embryonic diapause, nor would this be expected as it is a feature only of some of the kangaroos.

It is not likely that the breeding biology of the thylacine is much different from that of the Tasmanian devil in its main features, with a restricted breeding season and some out-of-phase breeding. Mating probably takes place in December or thereabouts since the young would then be born in January and reach the pup or half-grown stage by June–September as shown in Fig. 5.2, and start an independent life by early summer.

Late stage young

Several old-timers including H. Pearce have stated that when the young leave the pouch they run with the mother, hiding in shelters while she hunts. They also told me that at a later stage the young followed the hunting trail. The mother moved with her family from night to night and all of them hid for the day in any convenient spot. This statement, remarkable because it disagrees with many other authors who have said that the young were left in a den, was substantiated by Meredith (1881) who lived on the east coast near Swansea for many years and was in a position to gather information first-hand from trappers who, like Pearce, were experienced in the ways of thylacines.

Pearce's theory is supported by the number of female thylacines which were caught with cubs either for zoos (Chapter 4) or for bounty. If the cubs were left in a lair while the mother hunted they would not be taken with her when she was caught, but if running with her it is likely they would stay with her if she was snared.

The young of the Tasmanian devil show a very similar behaviour after quitting the pouch. The devils have paws which enable them to cling to the back fur of the mother and in this way they accompany her on the nocturnal feeding activities. At a later stage, they run beside or after her. I caught a mother in a trap at Cape Portland and the four young were climbing around the outside of the cage trying to get in. It is not unlikely that thylacines would have the same sort of behaviour, except that their young would not be capable of clinging to the mother after pouch quittal.

As early as 1852, West observed that thylacines formed family units, and several trappers told me that the young ran with the parents for a time and that 'packs' of up to six individuals had been seen. The young may stay with the parents for some months, probably until the next breeding season. The fact that the young were seen with both parents is unusual, as the formation of a family unit implies that some degree, however temporary, of pair bond exists between the parents. If thylacines form this bond, it may act against the rehabilitation of the species and would be likely to cause a slow recovery since the finding of a new mate in an already rare species would not be easy.

There is no information on the growth of the juveniles nor on the age of maturity.

Movements

We have no direct observations on the movements of thylacines and we do not know whether they were sedentary or had a territory or home range, or even if they were vagrants, wandering as the spirit moved them. However, some evidence collected from the Woolnorth station diaries as well as from the trappers themselves gives some clues to their habits.

In the diaries we read of thylacine hunts and the evidence indicates fairly strongly that the animal made little attempt to move out of the area even in the face of persecution (Chapter 6). This would suggest that they certainly had a home range, if not a territory. This view was supported by Mr A. Youd of Deloraine, who trapped in the Lake Adelaide–Golden Valley area, when he told me that 'once you found where they lived then all you had to do was to stick at it until you caught them'. A similar viewpoint was suggested by Wilf Batty in his account of the shooting of a thylacine at Mawbanna in 1930 when he made the point that the animal had been in the area for some time prior to the shooting. This also would support the home range concept.

What may appear to be a contrary view was expressed to me by the last tiger man on Woolnorth, George Wainwright, who believed that thylacines moved up and down the coast feeding on the abundant game living in this excellent habitat. The tiger man lived at Mt Cameron West and caught most of his thylacines on the coastal runs, and Wainwright believed that he caught them as they moved along the coast. I contend that it is possible that, with the catching and removal of thylacines from the coastal runs, other animals from sparser areas would tend to move into this lush habitat, thus giving an impression of constant movement.

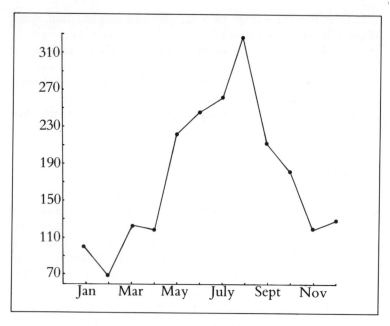

Fig. 5.3   Catch of thylacines each month of the bounty period. Based on the government bounty claims, Van Diemen's Land Company records, and the Hamilton Municipality 'Tiger Book'.

Most of the bounty kills at Woolnorth took place in June and July (Fig. 5.3) and this follows the pattern for the rest of Tasmania, although thylacines were taken in every month of the year. The seasonal nature of the catch led Malley (1973) to postulate that thylacines moved to the coast in winter as it offered a warmer environment and a better breeding area. He then went on to suggest that traditionally thylacines moved up the coast but in later times agricultural activity around the coast forced the animals to move through the thick inland forests.

It would be most unusual for any mammal to alter its traditional migratory route and there is no evidence that seasonal migrations took place. Although Malley's suggestion does not fit in with the idea of a home range, nevertheless his view is interesting and we know so little about the animal in its natural wild state that he might well be right.

From the bounty records we are able to identify the month of the year when the animals were killed as the clerks usually gave this information in the ledgers when making the entry for the claims, more being taken during the winter than in summer. From 1888 to 1908 about seventy thylacines were caught per month over the state during the period November to March, this rising to over 100 per month from May to October, peaking in July–August (Fig. 5.3).

The wintertime peak of kills is mainly related to the seasonal work of the snarers, although there is some evidence from Woolnorth (Chapter 6) indicating that thylacines were more active in the winter months. Toward the end of last century the fur industry was very important and most of the highland shepherds turned to snaring whenever the sheep had been returned to the lower winter pastures, as the high price for winter skins was a powerful financial incentive. The killing of wallaby and possum still takes place in the winter months but the industry has declined and snaring is no longer permitted, being replaced by shooting.

Some of the old records refer to thylacines moving down from the mountains in wintertime and returning in late spring and summer for breeding, but such little evidence as is available does not support this view. It is atypical of other Tasmanian mammals, none of which undertakes such a migration.

Activity

It is generally agreed that thylacines are nocturnal in their activity but nevertheless about 47 per cent of thylacine sightings have taken place during daylight hours (S.J. Smith 1980). Most of the species which form the food of the thylacine become active in the bush in the early evening, grooming themselves and moving to open areas to nibble at the grass, and some are still to be seen grazing or sitting around sunbathing and dozing well after sunrise. There is no reason to assume that thylacines would not be active at these times and Sawley (1980) states firmly that 'thylacines commence hunting just at dusk at all times'.

Thylacine sightings in the daytime have been reported by

both tourists and people experienced in the ways of animals in our bush. Jessie and Leo Luckman told me that while walking in the south west they were followed by a thylacine for several days, and the late H.R. Reynolds was followed by a presistent individual for several days when he was packing supplies from Tyenna down the Port Davey track in the early 1900s. Reynolds believed that the thylacine was attracted by the smell of the bacon which formed part of his load.

Renshaw (1938) observed that thylacines in the New York Zoo spent most of the day basking in the sun and also that they had poor vision in sunlight. The behaviour of captive animals is not necessarily the same as wild animals, but they too have been recorded as basking in the sun. The claim that they have poor vision by day is unlikely to be correct and is not supported by other writers.

Hunting and killing

Every predator must gain its food by the hunting and killing of prey and in this the thylacine is faced with the problem of catching prey which is sometimes larger and often faster than itself. Since the thylacine is a rather slow runner and so cannot run down its prey, it has either to stalk its prey and then jump out upon it like a leopard or else to hunt in a pack using co-operative hunting techniques like the wolves. There is no evidence that thylacines used any type of pack hunting in catching their prey.

From its similarity in appearance to the wolves and dogs, it might be expected that the thylacine would be a fast runner, but all of the old trappers were unanimous in stating that they did not have enough speed to do this in a straight chase, but they told of two other ways used by the thylacine to catch its prey. It is known that a wallaby when being hunted runs in a wide circle and a thylacine when chasing one would cut across along the chord of the circle and grab the animal as it went past. Thylacines were very presistent runners and could lope after their prey until the animal finally collapsed with exhaustion. Both these methods sound quite feasible and would be possible hunting techniques, but all trappers agreed that no co-operative hunting occurred.

The larger Dasyures show different hunting techniques ranging from the essentially scavenging devil to the three different predatory methods used by the other species, the native cat, the tiger cat, and the thylacine. The smallest of the three, the native cat, is quick and active and catches rats and mice by a rapid snap directed at the base of the head, killing the prey by a quick shake, thus dislocating the neck. The larger tiger cat is somewhat slower moving and kills rats by catching them and strangling them, probably by lying in wait and leaping out on its prey. Both the tiger cat and the thylacine have adapted their methods to enable them to catch prey larger and faster than themselves. The Tasmanian devil is slow moving, rather clumsy, and mangles its prey before finally killing it (Buchmann & Guiler 1977), but it probably very rarely acts as a predator being a very inefficient killer. Devils will kill disabled animals such as cast sheep (sheep on their backs, unable to rise) and have been known to play merry hell in a pen of chooks, but rather they are scavengers, eating remnants of animals and other food left lying around, and in fact they are very efficient in cleaning up the bush.

I do not know of any written descriptions of the method used by thylacines to kill their prey but I have talked about it with former hunters who presented some degree of agreement on the topic. H. Pearce told me that 'they hunt by lying in wait for their prey and then jump out on it. Roo are killed by standing on them and biting through the short rib into the body cavity and ripping the rib cage open'. Thylacines are anatomically suited to this method of killing having an extraordinarily wide gape which could be used to seize the neck or chest of a wallaby and so crush it. This method was described also by F. Sawley who lived in north-west Tasmania in 1900–1930, and he stated that the liver and heart were eaten (Sawley 1980).

Pearce's story must carry some weight and although his description of hunting and killing is somewhat different from the general views, the 'lying in wait' would fit in with the final phase of the chase after the pursuer had cut across along the chord of the circle. Pearce agreed that thylacines were slow and could not run down their prey.

A few trappers believed that one thylacine drove the prey

The enormous gape of a thylacine enables it to seize a wallaby by the throat or chest.

towards its mate, which jumped out of hiding and killed it (G. Stevenson, quoted in Hayes 1972). This implies a degree of co-operative hunting which was not reported by other observers, but it, like Pearce's account, can be reconciled with the cutting across in the final phase of the chase.

Having killed the prey it was then eaten. Many of the trappers spoke of a liking for the vascular tissues (G. Smith 1909; Hickman 1955; Saunders 1972; Sawley 1980). An interesting addition was made by Sawley to the effect that 'the tail was eaten about two inches from the anus to provide roughage'. The heart, lungs, kidneys, and liver were consumed together with some of the meat from the inside of the ham but the remainder of the carcase would be left and the devils would devour it.

One summer day in January 1952 about 10 km north of the Pieman River I came across a dead wallaby which had been killed in just this fashion, the corpse being still warm. Perhaps the thylacine had been disturbed at its meal. In 1976 about 16 km from that site I took a photograph of a pademelon which had been killed by having its throat eaten out. As with the earlier kill, only those parts mentioned had been eaten and there was no sign of other bites or of a struggle.

This killing method is quite different from that used by dogs, which characteristically 'worry' their prey. Tiger cats enter through the anus or belly, while devils are not selective and will start to eat injured or dead animals at any part that happens to be convenient.

In captivity the thylacines ate quite different food and would eat almost anything presented to them. Gunn (1850) kept some in captivity and fed them on mutton, but they preferred the parts containing bones and did not eat the vascular tissues. Thylacines kept in captivity by Rowe in 1913 ate bones as well as meat of wallabies (Rowe, quoted in S.J. Smith 1980), while the London Zoo animals were fed on rabbits and also caught pigeons for themselves, about which more later in this chapter.

There is general agreement that in the wild state thylacines kill their own food and never return to a kill (Sawley 1980). Carcases of sheep dosed with strychnine in the hope of poisoning thylacines were never eaten by them.

Animals in captivity, if they are going to survive, have to eat what is presented to them, rather like being in a boarding school, and this gives no indication of their feeding habits in the wild.

Vocalizations

Vocalization in the marsupials does not reach the high degree of development found in many eutherian mammals, and apart from agonistic encounters marsupials tend to be quiet animals. It is apparent when watching a marsupial make a loud noise that it is accomplished only as a result of consider-able effort and sound production does not appear to be easy.

It is not known exactly how many noises are made by thy-lacines or the circumstances in which they make them. Le Souef (1926) recorded a coughing sound and trappers told me of the same sort of noise but it is not clear what this call means.

A yapping sound likened to that 'of a dog barking but quite distinctive' (Nicholls 1960) is used apparently when hunting. The noise is a yap-yap or a nasal yaff-yaff according to Smith (1980, quoting other sources). On one occasion in 1961 I heard a call which corresponded closely to these descriptions. The yap-yap was high pitched with the second yap being lower in pitch and following the first very quickly, almost as an echo.

When thylacines are irritated they utter a low growl and when excited the inspiration is rapid and accompanied by a harsh hissing noise and this is used as a warning (Harrison, quoted in Bell 1967)

Locomotion

The gait is a trot with a distance between impressions of any one foot being about 70–75 cm though this varies with the size of the animal. A thylacine seen on the west coast in January 1970 was noted to be trotting like a trotting horse (D. Whayman pers. comm.).

There is no information on the running gait, though Le Souef & Burrel (1927) stated that when pressed thylacines break into a shambling canter. This resembles the running gait of the Tasmanian devil. A thylacine which broke out of a snare was noted to 'bound like a lion' (J. Le Fevre 1953). There are a number of early literature references to thylacines hopping like a kangaroo, especially when under a hard chase (Melville 1833; Martin 1836; Parker 1833), and Green (1974) quotes trappers as saying that thylacines take several kangaroo-like hops before assuming a normal gait. This may be the 'bound like a lion' of Le Fevre.

The bound is one of the most persistent legends which has sprung up around the thylacine but, apart from a few bounds in the early stages of a chase I doubt very much whether the animal could hop very far. We have seen that there was nothing abnormal in the anatomy of the hind limb (Chapter 3) and Barnett (1970) showed that there was a resemblance between the convexity of the transversely placed thylacine calcaneus and that of the kangaroos. This feature is found both in cursorial as well as saltatorial forms and so cannot be considered as convincing evidence for bipedal locomotion. The work of Cunningham (1882) did not show any unusual features of the hind limb musculature, nervous system, or osteology which would allow the thylacine to indulge in bipedal frolics. Furthermore the tail is rigid and this surely would get in the way of any hopping action.

The study by Keast (1982) showed that the limb proportions of the metatarsal segments of the thylacine are shorter in relation to the total limb length than in the wolf and there is no anatomical suggestion that thylacine hind limbs are in any way adapted for bounding. The legend may have arisen because the thylacine has been seen to raise itself on its hind legs in tall grass for a few moments to have a look around, and dogs do this also under similar conditions.

Thylacines have been reported to be able to leap like a cat, one person assuring me that they could jump a 6-metre high woodheap from a standing start without touching the wood. The thylacine shot by Wilf Batty had been in the area for several days and had been seen in an adjoining paddock where it had jumped a fence 145 cm (4 feet 9 inches) in height touching only the top strand with its forefeet. This would appear to be within the capabilities of the species whereas the woodpile tale is hard to accept.

Thylacines are fairly agile, Gunn (1863) recording them as jumping from one crossbeam of their compound to another. A similar incident is recorded by S.J. Smith (1980).

Another legend which is heard is that thylacines cannot swim and are fearful of the water. This is highly unlikely and the earliest evidence against this comes from the Tasmanian Aborigines who said they swam very strongly (Milligan 1853). Maybe this is where Swainson (1846) got the notion that they were an aquatic species which caught fish. Other instances of observations on swimming are given by S.J. Smith (1980). However, Mudie (1829) was carrying the matter a bit far when he suggested that thylacines frequented the seashore where they ate carrion and probably swam after fish.

The true nature of thylacine locomotion and vocalizations will not be known until observations can be carried out on animals living in the wild.

Parasites and diseases

Thylacines undoubtedly were hosts to a wide range of parasites, both internal and external, but we know little of them nowadays. Old movie film shows a thylacine scratching itself vigorously, perhaps due to the nibbling of the flea *Uropsylla tasmanica*. This flea was recorded from a thylacine in Launceston in 1879 (Dunnet & Mardon 1974). This species of flea is widespread in Tasmania and is characteristic of the Dasyures. Pearse (1981) found that tanned thylacine skins contained burrowing larvae of a related species of flea.

Ranson (1905) and Sprent (1971, 1972) described two internal parasites both of which were acquired in captivity. Ranson's parasites were tapeworm cysts found in large numbers in the involuntary and heart muscles of a captive animal and he believed that these were an infection from other zoo animals. A nematode worm, *Cotylascaris thylacini*, was described by Sprent (1971) from the last thylacine in the London Zoo. He discovered subsequently that the worms were in fact *Ascaridea columbae* (Sprent 1972), of which the usual host is the pigeon. Apparently, the unfortunate pigeon had ventured into the thylacine's cage and had been caught and eaten by the no doubt delighted thylacine. Munday & Green (1972) record that Mawson found a species of nematode worm, *Nicollina* sp., from a thylacine.

In an earlier chapter I have referred to the possibility of a distemper-like disease which may have reduced the thylacine population as well as that of other Dasyures. Nothing is known of this disease, it was not distemper nor has it ever been suggested that it was, but I have often been misquoted as saying so.

Field signs

The chances of actually seeing a thylacine in the bush are very remote and we have to rely on field signs such as footprints, droppings, and kills as evidence of their existence, and the interpretation of these signs requires much study and skill.

Footprints are the evidence most frequently found. They are rarely complete or clearly impressed into the substratum and may even be distorted by leaves, twigs, or stones and

Probable thylacine dropping, Three Sticks Run, Woolnorth, November 1961. The droppings contained Hereford calf fur; the length of the knife is 22·5 cm.

gravel. The illustrations issued by the National Parks and Wildlife Service are of help but a plaster cast should be made and the print sent for expert identification.

Since the droppings of thylacines resemble those of the Tasmanian devil and both contain a lot of hair as well as bone fragments, this is not a reliable sign of the presence of a thylacine. The size of the dropping is not diagnostic although those of the thylacine tend to be large and some have a peculiar twist. The notion of a twist in the scat prompted Mrs Marjorie Dainton, formerly of Rossarden, to write this little poem and enclose it when she was sending me some possible thylacine scats:

> It came by post
> 'Round about noon
> When crowds of people
> Were in the room.
>
> We watched and listened
> And then we heard

'Gawd, what's this?'
Was it a bird?
'No' he said 'don't be absurd
It's only a silly twisted turd'.

Returning to more prosaic matters, the droppings contain
hairs derived mainly from the animal's last meal and I see
little point in indulging in the intense labour of sorting
through masses of hair in the hope of finding a few which
might come from the animal grooming itself. I spent three
days once going through one scat which yielded only wallaby hairs. A biochemical technique is currently being investigated by the National Parks and Wildlife Service which
is hoping to be able to identify the scat using the presence of
bile salts. Each species may have its own characteristic com-
position of the bile and it may be possible to identify the scat
positively from a chromatographic examination of a small
part of the material. This analysis showed some early prom-
ise but the separation of the salts has not yet produced a pattern which can be used to identify a species. If successful, this
technique would make the identification of any species possible from a scat.

A thylacine kill is quite characteristic but unfortunately the
carcase is not around for long before devils devour it completely and leave no trace.

Thylacines and agriculture

Before leaving the biology of the thylacine it is desirable to
consider the possible consequences of certain agricultural
practices on the thylacine.

Poisoning is the most commonly used means of controlling kangaroo and wallaby populations where necessary. The
poisons are mixed with a base, either dry or fresh, usually
bran, pollard, carrots, or apples. The most commonly used
poisons are strychnine and 1080 (sodium monofluoroacet-
ate). As thylacines are not known to eat any of these vege-
table materials the poisoned baits present no danger to them,
and their predatory and non-scavenging habit of eating only
freshly killed meat protects them from secondary poisoning
due to eating poisoned animals.

The two poisons act in different ways on the target species, strychnine killing the animal quickly close to the bait site, whereas 1080 does not kill quickly and the unfortunate animal often moves some distance from the bait line. It is possible that a thylacine would kill and eat such an animal. It is not known whether thylacines are tolerant or otherwise to 1080. Their relative, the Tasmanian devil, is known to have a considerable tolerance for the substance and is unlikely to die from secondary poisoning after eating a carcase. If thylacines have a high tolerance for 1080 then we can expect that secondary poisoning is unlikely.

The clearing of land and consequent alteration of habitat affects all of the species living in the cleared area and also those in the surrounding bush because of the invasion of their range by the displaced animals. In the initial stage of clearing when the trees are felled and the site burned the effect is drastic as no food is available for the grazers. However, when grass or crops are planted the picture changes and the paddocks provide lush grazing. For the thylacine the increased availability of prey would be more than offset by the absence of shelter in the cleared area. The land clearance which has taken place since 1804 has resulted in large areas of Tasmania being turned into unsuitable habitat for thylacines.

The introduction of alien species to Tasmania provided thylacines with a new source of food as we have seen how readily they turned to killing sheep. Some trappers claimed that the introduction of rabbits had a serious effect on thylacines because their soft fur formed balls in the thylacines' intestines causing a blockage of the gut. No opinion can be formed on these views as there is no evidence that thylacines could catch rabbits and, if they did so, that they ate the whole animal, and no conclusions can be drawn about the passage of the rabbit fur through the thylacine's gut.

The constant grazing by sheep and rabbits over the years has considerably altered the plant environment and has reinforced the efforts of farmers and foresters to reduce habitat which would be suitable for thylacines.

6

The Van Diemen's Land Company and the thylacine

The early exploration of Tasmania and the growth of agriculture were assisted in no small way by a system of land purchase and land grants to individual settlers or to companies. The Van Diemen's Land Company was established in London in 1824 to develop farming and other interests in Tasmania. Other chartered companies had succeeded in other parts of the world, notably in Africa and India, and this encouraged investment in the VDL enterprise (Meston 1958).

The VDL Company was granted 500 000 acres of land on 9 November 1825 and searching for the best locations started in 1826, but as 'no good land' was left it was allowed to select 250 000 acres in north-west Tasmania at two shillings and sixpence per acre. Governor Arthur did not allow the Company to have this holding in one large allotment and refused it some better land at Port Sorell. The Company prospected large areas of the rugged country behind the north-west coast and eventually received 150 000 acres at Woolnorth, 20 000 acres at Circular Head, another 10 000 each at Hampshire and Middlesex Plains, with a further 150 000 acres at Surrey Hills (Fig. 6.1). The offshore islands were included and these con-

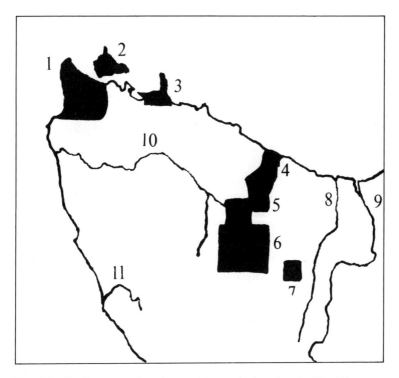

Fig. 6.1   Outline map of north-west Tasmania showing the Van Die-
men's Land Company properties: 1 Woolnorth; 2 Robbins Is.; 3 Circular
Head; 4 Emu Bay; 5 Hampshire Hills; 6 Surrey Hills; 7 Middlesex
Plains; 8 Forth River; 9 Mersey River; 10 Arthur River; 11 Pieman
River.

tained another 10000 acres. The Company was very active
and by 1834 no fewer than 400 persons were employed, of
whom about 200 were assigned convicts (West 1852).

A large tract of land in north-west Tasmania extending
south from Emu Bay, now the site of the town of Burnie,
formed part of the grant together with Woolnorth Point at
the north-western extremity. The early settlement, the land
clearance programmes, development of industries and pastoral interests all were part of the Company's activities.
These soon extended from agriculture into railway owner-
ship, the Emu Bay Railway being a VDL Company venture,
while brickworks and logging and sawmills were owned

by or associated with the Company, together with food-processing activities such as flour mills and a butter factory. Woolnorth is now the Company's sole remaining property and all of the other interests have been sold.

We are not concerned with the effect of all this activity on the exploration and development of the country in general but only with its significance to areas where thylacines lived.

The Company was highly organized, and detailed reports of its activities were collated, condensed, and then sent to London. At each station the Company ran a store and credit was given to staff for purchases which were duly recorded in the individual account books. These books not only provide details of the requirements of the workers but record the killing of thylacines since credit for them was given at the store. The staff were helped with cash advances from time to time and we discover that, for example, E. Childers was loaned £1 'for the purchase of articles required on his passage to London'—hardly generous, but the monthly wage was only £6.5.0.

All of the day-to-day work duties of the staff were recorded in the station diaries, which were maintained at each property. In addition a monthly stock return was made at each station, and these two sources provide valuable information on the thylacine. Diary entries were made on every day including Sundays during the year except for Christmas Day when the entry was 'Blank Day'.

We read of vagabonds stealing sheep and being chased from the property and that 'natives raided the sheep', and hunts for escaped convicts are mentioned. Probably all a welcome break in the dull routine of station life.

The diaries for Woolnorth are reasonably complete but with gaps in the early days and in the 1850s and 1860s, but those of the other properties have disappeared except for a few years at Surrey Hills. These records are the only strictly factual accounts of thylacines during the pre-bounty days and are devoid of any hysterical exaggeration or distortion. They give us a picture, however dim, of how thylacines lived and their effects on a large farming community.

We must not think of these stations as primitive outback communities with no facilities. Although not as big as Woolnorth, the station at Surrey Hills must have been quite an

Woolnorth homestead.

impressive place, having a staff in 1844 of three officers of the Company, one of whom was a surgeon, eleven mechanics and labourers, and two assigned servants. The Woolnorth staff was about twice that size with three times as many assigned servants.

Woolnorth

Many Tasmanians have never heard of Woolnorth or they know of it only vaguely as being in the west somewhere; its name, which has an air of mystery and romance, is found on few maps. The concept of a large farm covering all one end of the island allows the imagination to run riot with ideas of what goes on there. The Company has always kept itself rather aloof from local affairs, and its geographical position in the far north west has made it somewhat inaccessible, especially in winter when wet muddy conditions are encountered everywhere. It has been comparatively easy to fence the station off from visitors, while shooters have never been encouraged, so a legend of mystery has built up.

The present Woolnorth property is smaller than the original grant, now being about 100 000 acres. The early homestead was located at the site of the present buildings and was supplied largely through a port which was established near Woolnorth Point. Indeed, during my first visits to the property in the winter of 1958 the road was so bad that I wondered why they ever gave up the notion of supply by ship.

The homestead in those early days presented an entirely different social structure, perhaps more feudal in nature than would be acceptable nowadays. There was little contact with the outside world except for the arrival of ketches from Stanley with supplies. The isolation meant that local events became of great importance in the lives of the people, and the centre of activities was the stock and farming. Everything that happened was recorded in the station diary in some detail.

One of the first activities at Woolnorth was clearing the land around the selected homestead site and this was laid out in a series of paddocks. The remainder of the property was left more or less in a natural state and divided up into a series of runs. The paddocks were used for close grazing or for growing crops and each paddock was named. These names were used in the diary and are still used today so we have full details of where events took place and it is possible to trace the locations on the property.

A major task was fencing, some of the material for this was imported from Burnie and the remainder cut on the property. One type of fence, built from wooden posts and wooden palings, was time consuming to construct as each fence post had to have three or four slots cut through it to hold the horizontal wooden palings, each end of which had to be tapered to fit into the slots. Another type of fence used on the property consisted of posts and wire which supported vertical palings, the wire being twisted around each paling leaving about 3-inch (8 cm) gaps between palings.

Work continued every day, in fair weather or in high winds and heavy rain, and something of the atmosphere of wintertime misery comes through in the brief entries in the diary. For instance, it is recorded continuously for about two weeks during one winter that 'the boy weeded the turnips'. The weather always was important enough to be noted every day and the constant wet conditions must have been difficult to endure. The lower parts of the property, mainly the paddocks, became waterlogged in winter and largely impassable to any form of transport other than walking. The drains of the flat country filled with water in winter and perhaps the ultimate of misery must have been in 1885 when all hands were 'draining' in all weathers from 6 May until 19 May.

Thylacines on Woolnorth

The Company must have had some trouble from thylacines killing sheep at an early stage in the development of the property, but I have found no comments or statistics about this in the few available records, and in fact, the only statistics are to the contrary. The tables of stock increases at Woolnorth 1832–34 give no losses due to thylacines nor do they pass comment on this topic although they record all sorts of other calamities, including between 80 and 150 'accidents' per annum. By 1839 sheep were being killed by thylacines although the number is not known, but by 1843 the losses had increased greatly (Table 6.1) and something had to be done about it.

The first 'tiger man' was appointed at Woolnorth in 1838, being given his keep and a reward for each thylacine killed. He also sold the skins of other animals which he caught. I do not know if he was paid a wage as well. He travelled from Mt Cameron West to Woolnorth every month or so with his skins, collected his supplies and went back a few days later. All of this was duly recorded in the diary but the number of thylacines which he killed has been lost.

The method of catching thylacines and other game was by snaring, usually by the use of 'necker' snares which consist of a noose of wire or hemp placed around a hole in a fence or a constriction in an animal trail in such a way as to strangle the animal when it became caught in the noose. The removal of one of the palings from the wire fences allowed game to pass through and at the same time provided an ideal snaring site. Segments of this old fencing could still be seen in 1960–62, particularly on the Three Sticks run in the shelter of tea-trees where we used the gaps as positions for our cameras. It is doubtful if treadle and springer type snares were used, largely

TABLE 6.1   Sheep lost by predation, Woolnorth, May 1839 to September 1850

Date	Sheep loss	Predator	Date	Sheep loss	Predator
1839 May	1	Thylacine	1845 July	28	'Vermin'
1840 April	1	Thylacine	Oct.	7	'Vermin'
May	1	Thylacine	Nov.	38	Hyaenas, etc.
Sept.	2	Thylacines	1846 April	13	Hyaenas
1841 March	2	'Vermin'	July	14	'Vermin'
July	2	Thylacines	Oct.	36	'Vermin'
Aug.	4	Thylacines	1847 Jan.	18	'Vermin'
1843 March	4	Dogs	Feb.	8	Hyaenas
April	21	Dogs			and dogs
June[a]	69	Dogs	Mar.	10	'Vermin'
	16	Thylacines	April	7	'Vermin'
July	24	'Vermin'	May	6	'Vermin'
Aug.[b]	47	'Vermin'	June[c]	54	Dogs
	130	Dogs	Sept.	67	Dogs
Sept.	19	'Vermin'	Dec.	52	Dogs
Oct.	65	'Vermin'	1848 March	30	Dogs
Nov.	66	'Vermin'	June	139	Dogs
1844 Feb.	3	'Vermin'	Sept.	94	Dogs
March	19	'Vermin'	1849 June	305	'Vermin'
April	2	'Vermin'	Sept.	216	'Vermin'
1845 April	23	'Vermin'			and weather
May	35	'Vermin'	1850 Sept.	42	'Vermin'
June	10	'Vermin'			

[a] The 69 killed by dogs occurred during a slaughter of 64 Leicester lambs.
[b] Of the 130 killed by dogs no fewer than 78 were driven into the sea.
[c] From June 1847 the stock return seems to have been sent in quarterly.

Source: Compiled from the monthly stock returns of the Van Diemen's Land Company for the Woolnorth station.

because there were few good springers to be had for miles around. They were so scarce that we carted ours from Tarraleah.

Thylacines caught in necker snares may well have been dead or very battered when found although a few were healthy enough to be sent to zoos in later days, the last 'tiger man' G. Wainwright sending three to Hobart Zoo in 1913 or 1914. These were caught on Mt McGaw (Wainwright pers. comm.).

Although the 'tiger man' was based at Mt Cameron, he

operated all over the property as needs demanded. Thylacines were encountered over every run and paddock and whenever one was sighted near the station all hands were turned out to scare it off or shoot it. Thus we read on 14 July 1879 'all hands trying to shift a tiger at Studland Bay', and on 20 February 1898 'tiger scaring at Three Sticks and Studland Bay' and again on 14 May 1904 'all hands hunted a tiger'. Knowing that a tiger was around on the property seems to have sent the station into a frenzy of activity to get rid of it. On 29 April 1901 a tiger 'was being chased in the Forest' while on 18 May the chase was at the inlet and at the forest. No success is mentioned and further scaring took place there on 11, 16, and 17 July 1901. Finally six thylacines were caught on the property at this time, one was 'dogged' on 17 July at Three Sticks, one was caught in a snare at Spink's Paddock on 18 July and two were snared at Valley Bay, another was caught at McCabe's Paddock on 19 July and yet another was snared there a day later. A bounty of £6 could be claimed on these, and as that amount represented a month's wages it was a rich haul.

These activities show that thylacines were numerous around the property, the six being caught within an area of 4–5 square miles (10–13 sq. km), and further, except for the Three Sticks caught in paddocks which are very close to, if not adjacent to, the homestead. Unfortunately the records do not state if the animals were male or female or whether they were adults or young.

Some hunts were not so successful, and we look at the year 1897 as an example. On 25 January attempts were made to shift a thylacine off the Welcome Heath by burning the scrub but all hands were back there on 2 and 5 May hunting an animal and the 8, 9, and 10 May were spent hunting a thylacine in the nearby forest and back to the heath again on 22 and 30 May, then to the forest again on 3, 4, and 5 June without any success. From this we can appreciate that catching a thylacine has never been an easy task. For the rest of the year regular hunting took place mostly at Studland Bay and Three Sticks but the quarry remained elusive. Details of these activities are given in Table 6.2.

TABLE 6.2   Details of thylacine hunts, Woolnorth, 1874–1914

1874—	21 Aug.	One tiger killed.
1875		No comments.
1876	4 Mar.	Tracked a tiger at Studland Bay.
1877		No comments
1878	13 Sept.	Trying to shift a tiger up the coast from Studland Bay.
1879	14 July	Trying to shift a tiger at Studland Bay.
1880	6 July	Chased at tiger from Studland Bay run.
1881		No comments.
1882		No comments other than collected skins and hides from the Mount.
1883		No comments.
1884		No comments.
1885		No comments.
1886	26 June	Snares set at Green Point. *a*
1887	3 Jan.	To Green Point snares.
	31 Jan.	To Green Point snares. *b*
	8 Apr.	All hands chasing tiger from Forest.
	23 June	To Studland Bay to shift a tiger.
	18 Aug.	Tiger in the Forest.
	19–20 Aug.	Snares set in the Forest.
	23 Aug.	One tiger located at the Mount.
	24 Aug.	Forest snares empty. Also empty on 30 Aug., 3 Sept., 7, 10, 15, and 28 Sept.
	5 Oct.	Tiger at Studland Bay.
	7 Oct.	Forest snares empty. Also empty on 11 Oct.
	10 Nov.	Mount snares empty. Also empty on 15 Nov.
1888	14 Mar.	Tiger man came home.
	23 May	Tiger man at Arthur River, snares not looked after.
	15 June	Trying to shift tiger out of Forest.
	16 June	Trying to shift tiger out of Forest.
	21 June	Looking after tiger on Swan Bay run.
	28 June	All hands at Studland Bay trying to shift tiger. Tracked him.
1889	26 Aug.	Tom went to the Mount to look after a tiger with his dogs.
	2 Nov.	Sent some men to hunt tiger out of Studland Bay run.
	11 Nov.	All hands in a.m. hunting a tiger out of the Forest. Set snares for a tiger on Saltwater Creek fence.
1890	6 Feb.	Tracks seen in Forest.
1890	27 May	Chasing a tiger in the Forest.
	17 July	Tiger at Studland Bay, at the Knolls.
1891	1 Aug.	Laid poison at Harcus for hunters' dogs.
1892	26 Aug.	Two men to Studland Bay to shift tiger.
1893		No comments.

1894	9 April	Davey caught a tiger in Park Paddock.
	28 Apr.	Went to Studland Bay to shift tiger.
	15 June	Some of the men went to Studland Bay to shift tiger.
1895	24 June	Tracked a tiger on Welcome Heath. Went with dogs in the afternoon to try to shift him.
	25 June	Looking after tiger on Welcome Heath.
1896		Diary no longer in existence.
1897	25 Jan.	Shifting tiger off Welcome by burning the scrub.
	30 Apr.	Track seen on Welcome Heath.
	2 May	Shifting Welcome Heath tiger.
	5 May	Shifting Welcome Heath tiger.
	8 May	Hunting tiger in the Forest.
	9 May	Hunting tiger in Forest.
	10 May	Hunting tiger in Forest.
	22 May	Hunting tiger on Heath.
	30 May	Hunting tiger on Heath.
	3 June	Tiger seen in the Forest.
	4 June	Tiger chase in the Forest, unsuccessful.
	5 June	Tiger chase in Forest.
	17 July	Tiger on Studland Bay and Three Sticks runs.
	5 Aug.	Tiger at work on Studland Bay run.
	6 Aug.	Went to shift tiger.
	14 Aug.	Still looking for Three Sticks run tiger.
	28 Aug.	Still looking for Three Sticks run tiger.
	7 Sept.	Still looking for Studland Bay tiger.
1898	20 Feb.	One tiger caught, no locality given.
	20 July	One tiger caught, McCabe's Paddock.
	31 Dec.	Snaring in the Forest.
1899	3 July	Saw two tigers at Swan Bay.
	6 July	Caught two tigers in Forest and Three Sticks.
	22 July	Tiger scaring on Three Sticks and Studland Bay.
	23 Nov.	One tiger caught, probably at the Mount. ᶜ
1900	24 Jan.	One tiger caught, locality not stated.
	8 Feb.	Tiger scaring at Three Sticks.
	10 Feb.	Tiger in snares, no locality.
	12 Feb.	One tiger caught, no locality.
	13 Feb.	One tiger caught, no locality.
	25 Mar.	One tiger caught, no locality.
	9 Apr.	All hands tiger scaring, Three Sticks to Studland Bay.
	23 May	One tiger caught at Bullock Paddock.
	13 June	Two tigers caught, probably at the Mount. ᶜ
	14 June	One tiger caught. ᶜ
	5 July	All hands tiger scaring on Three Sticks and Studland Bay.
	9 July	One tiger caught, no locality. ᶜ
	19 July	One tiger caught, the Mount beach.

	12 Aug.	Tiger scaring in Forest.
	13 Aug.	Tiger scaring on Three Sticks and Studland Bay.
	18 Aug.	One tiger caught, no locality. [c]
	19 Aug.	One tiger caught, no locality. [c]
	2 Sept.	Tiger scaring at Studland Bay.
	6 Sept.	Two tigers caught, Valley Bay.
	9 Sept.	Three tigers caught, no locality. [c].
	10 Sept.	One tiger caught, no locality. [c]
1901	8 Jan.	Tiger seen in Western Mile Marsh.
	9 Jan.	Snares set on Welcome Heath and in Forest.
	11 Feb.	Tiger scaring at Forest and Welcome Heath.
	29 Apr.	Tiger scaring in Forest.
	2 May	Tiger snaring at Studland Bay. One tiger caught, no locality. [c]
	16 May	Tiger snaring on Welcome Heath, Inlet and Forest.
	18 May	Tiger scaring and snaring, Forest and Inlet.
	25 May	Snaring in Forest.
	30 May	Scaring and snaring in Forest and Inlet.
	1 June	Snaring at Three Sticks, Bluff, Welcome Heath and Inlet.
	28 June	Two tigers caught, Studland Bay. [c]
	11 July	Scaring tigers in Forest and Inlet.
	13 July	Snaring on Three Sticks and Bluff.
	16 July	Scaring tigers in Forest and Inlet.
	17 July	One tiger 'dogged' on Three Sticks. Scaring and snaring at Forest and Inlet.
	18 July	One tiger in Spinks' Paddock and two in Valley Bay snares.
	19 July	One tiger caught in McCabe's Paddock.
	20 July	One tiger caught in McCabe's Paddock.
	31 Dec.	Snaring in the Forest.
1902	10 Mar.	One tiger caught, no locality. [c]
	28 Mar.	Tiger scaring on Welcome Heath.
	24 May	Tiger scaring on Three Sticks.
	26 May	Tiger scaring and snare setting on Forest.
	10 June	One tiger caught, Studland Bay. [c]
	16 June	Snaring in Forest.
	27 June	One tiger caught, no locality. [c]
1903	20 Apr.	Dog caught in a tiger snare. [d]
	30 Apr.	T. Well on Three Sticks run, tiger hunting.
	30 May	Four men on Three Sticks run, tiger hunting.
	15 Sept.	Two tigers caught round lambs and ewes. [e]
1904	14 May	All hands to tiger hunt.
1905	19 May	Snares set on Mt Cameron.
	16 June	Snares on Mt Cameron examined. [b]
	14 July	Snares examined again. [b]
1906	19 May	Snares examined, no locality. Also examined on 23 May.

	21 June	Snares examined.[b]
	30 Sept.	One tiger caught at Studland Bay.
1907	13 Aug.	Snares at Studland Bay examined.
1908		No activity reported.
1909	7 July	Snaring at Studland Bay.
	11 July	Snares examined, also on 13 July.
1910	26 May	Studland Bay snares examined.
1911		No further activity reported from this year.

[a] It seems that when snares were set at unusual places this was duly recorded. Snares always were set from Mount Cameron to Three Sticks.

[b] Note the long gaps between examinations, any animal would certainly die in this time.

[c] Caught by Wainwright. These animals were presumably caught either at Mt Cameron or Studland Bay since Wainwright worked in those places and stated that he caught nearly all of his seventeen tigers there.

[d] I wonder how many dogs were killed in this fashion. The wild dog population certainly would have suffered casualties, all of which would have died before being removed from the snares.

[e] This is the only mention of presence of presence of thylacines round the sheep.

Source: Based on the station diaries.

A chase did not necessarily lead to the animal moving away from the area and seldom to its being caught. I do not know why so much time was spent in chasing these animals as they could only flee to another part of the property where they would still be a nuisance, and I rather suspect that the activity and excitement came as a welcome relief from the dull routine and isolation of the farm.

It is evident that these thylacines were reluctant to move away from their haunts, and this could suggest that they establish themselves in a home range as do other carnivores. The only way to get rid of them was to kill them.

Most thylacines were caught in the Mt Cameron West area where the tiger man operated most of his time, but numbers were taken at Studland Bay as well as on the adjacent Three Sticks (Table 6.3). Not many were caught on the Harcus or Welcome Heaths as in those days they were covered by thick scrub and were not the open paddocks and cleared areas we see today.

Study of the tables shows that most of the thylacines were caught on the coastal runs, only 15 per cent being caught elsewhere. This would be in accordance with the distribution of their prey which abounds on the consolidated sandhill

TABLE 6.3   Thylacines caught, and incidents reported, Woolnorth, 1874–1911

Locality	Killed	Incidents	Description
Mt Cameron	2 (9)	2	coastal
Studland Bay	3 (8)	24	coastal
Three Sticks	2	13	coastal
Valley Bay	4	0	coastal
Forest	1	24	inland
McCabe's	3	0	homestead
Spinks'	1	0	homestead
Bullock	1	0	homestead
No details	7	–	–
Green Point		3	coastal
Welcome Heath		13	inland
Swan Bay		2	coastal
W. Mile Marsh		1	inland
Inlet		6	coastal
Bluff		3	coastal
Park Paddock	1		
	25 (17)	91	
	36 Coastal	53 Coastal	
	6 inland	38 inland	

Note: The number in brackets refers to Wainwright's catch (see footnote *c*, Table 6.2).

country adjoining the sea. The thylacines presented for bounty in the north-west coastal towns would have been caught in the coastal hinterland, whereas inland centres such as Waratah did not record a high number of bounty payments (Appendix I). The large number presented at Stanley probably came off Woolnorth, since the local itinerant carrier as he moved around the country with his cart and goods collected the carcases from the trappers, paid the bounty from his own pocket, and later reimbursed himself when he presented the carcases at Stanley for the government bounty (Wainwright pers. comm.).

A suggestion was made by Malley (1973) to the effect that thylacines moved in a circular pattern up the coast to Woolnorth Point and then along the north coast in an easterly direction as far as Smithton and then south through the bush

Studland Bay, where so many thylacines were caught by the VDL Co. tiger men.

to the coast again. He may be right but the Woolnorth evidence presented in the tables suggests strongly that the thylacines probably were resident in an area and stayed as much as possible on the coastal runs. All of the seventeen thylacines caught by G. Wainwright were caught on the coast and the majority from the Mt Cameron cattle yards. Most carnivores do not indulge in such migrations and tend to have a home range or territory, and there is no evidence from the Woolnorth records that the thylacine does not conform with the others.

Thylacines were caught at Woolnorth during all months of the year but the highest catches were made in the winter months, and this was also the pattern elsewhere. We have already mentioned that most trapping takes place in the winter months when the pelts fetch prime prices and there is little risk of damage to the pelts from blowfly strike, and the fact that thylacines were caught at this time of the year is of some biological significance as it shows that they did not retreat to the hills and rugged areas for breeding as so many of the earlier writers tell us. Had such a movement taken place the animals would have had to move off the property to find rugged hilly country. The catches at Woolnorth at all times of the year contained both old and young animals.

The tiger men led lonely lives at Mt Cameron West. Their names have in many cases been lost to us but some have been perpetuated in the naming of the paddocks, for example Spink's Paddock. One of the first appointees was Th. Jackson (1839) and he was followed by J. Jackson who was appointed Senior Assistant in October that year, and at the same time W. Hudson was appointed to be stationed at Studland Bay. From this time on, the location of the shepherd was not mentioned in the diaries. Wainwright (pers. comm.) told me that he lived at Mt Cameron. Some of the tiger men were A. Walker (1850), B. Spinks (?), W. Forward (1877–81), J. Dowle (1881), C. Williams (1882–86), G. Wainwright senior (1890–July 1901), G. Wainwright (1901–14). After about 1909 there was no longer any need to employ a tiger man to trap thylacines.

In carrying out their job of protecting the Company's flocks, these men contributed toward the plight of the thylacine today, but they cannot be blamed for doing so, and from the records of their activities on the properties we have some most worthwhile data on this elusive animal.

Sheep losses
We have some early records of the sheep losses at Woolnorth and at Surrey hills (Tables 6.1 and 6.4), and it could justifiably be asked whether it was necessary to employ a tiger man to protect the sheep on VDL Company lands. The total sheep losses on Woolnorth are known for 1842–1848, when they were 1119, 1003, 899, 929, 302, 800, and 889 respectively, totalling 5941 sheep. The number of sheep on the station at the time was about 6800 so that the losses annually were in the order of 16–17 per cent, a figure which would be hard to sustain over a long period.

It is unfortunate that so many of the predations were recorded in the diary as being due to 'vermin' or 'hyaenas and dogs', but from the figures in Table 6.1 for the years 1839–50 it is clear that dogs were a much more serious predator than

TABLE 6.4   Monthly sheep losses at Woolnorth and Surrey Hills

	Jan.	Feb.	Mar.	Apr.	May	June	July	Aug.	Sept.	Oct.	Nov.	Dec.	Total
Thylacines													
Woolnorth	0	0	0	14	2	16	2	4	2	0	0	0	40
Surrey H.	6	38	3	5	4	5	3	14	16	5	0	1	100
Dogs													
Woolnorth	0	0	34	21	0	262	0	130	161	0	0	52	660
Surrey H.	10	2	124	35	8	27	4	22	6	1	5	9	253
Dogs or thylacines or 'vermin'													
Woolnorth	18	11	44	25	35	315	66	47	277	108	104	0	1050
Surrey H.	87	134	95	86	69	6	10	15	5	17	0	24	548

Source: Compiled from the monthly stock returns from Woolnorth (May 1839 to Sept. 1850) and Surrey Hills (June 1832 to December 1852).

TABLE 6.5   Sheep losses in various runs and paddocks at Woolnorth, 1888–95

	1888	1889	1890	1891	1892	1893	1894	1895	
Mt Cameron	27	34	18	24	24	17	26	27	197
Forest	0	27	21	0	16	7	11	32	114
Welcome Heath	0	2	2	0	6	0	2	5	17
Three Sticks	0	5	7	18	10	10	7	23	80
Studland Bay	6	13	4	2	2	0	3	6	36
'Paddocks'	21	7	3	13	3	18[a]	10	5	80
Victory Run	0	0	0	15	0	0	0	0	15
	54	88	55	72	61	52	59	98	539

[a] No less than 10 of these sheep were poisoned when a new cure for footrot was used.

thylacines. Although they varied from year to year, the sheep losses at Woolnorth from predation were only a small proportion of the total sheep losses.

It may not be significant but thylacines killed mostly merinos and not the five other breeds maintained on the property.

The appointment of a tiger man was made not so much to reduce the number of sheep killed by thylacines but more to control the losses by predation from both dogs and thylacines which constituted about one-third of the total sheep losses. It was easier to justify the appointment of an extra hand if a native species could be blamed for the trouble rather than the wild dogs for which there is always a certain amount of sentimental attachment.

Most of the sheep losses at Woolnorth occurred on the runs rather than in the paddocks (Table 6.5) and more sheep were killed during the winter months by both dogs and thylacines than in the summer. Thylacines would have to hunt for food more actively in winter than in summer when the young wallabies and pademelons are inexperienced and therefore readily accessible prey. The higher wintertime catch reflects also the activity of the trappers as well as that of the animals searching for food. Unfortunately we have no data on the number of dogs killed at any time.

Earlier work on other species, notably the potoroo and the Tasmanian devil (Guiler 1958, 1970) shows that catches of these species in the summer were much lower than in the

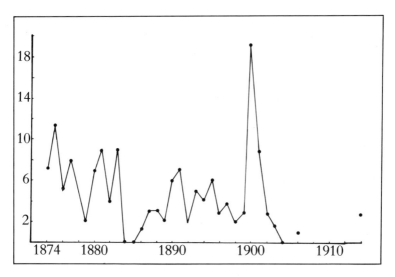

Fig. 6.2   Annual catch of thylacines at Woolnorth, 1874–1914.

winter and in the case of the Tasmanian devil yielded insuf-
ficient animals to make summer trapping worthwhile. Although some of this work continued for as long as thirteen
years, we were never able to explain the low summer activity. A similar behaviour pattern may occur in the thylacine,
and indeed, the data in the tables suggest this.

The catch of thylacines at Woolnorth peaked in 1900
whereupon a decline started which did not cease and only
sporadic catches were made after 1903 (Fig. 6.2). The nineteen thylacines killed in 1900 made a bumper year and the
population never recovered. In the later days of thylacine
hunting at Woolnorth it became clear in the diaries that enthusiasm for the job was not as great as before, probably due
to lack of results. The snares were left unattended for quite
long periods, the inference being that it was not worth going
around looking for the occasional animal. Thus we read '19
May 1905—Snares set at Mt Cameron' and not until 16 June
do we read 'Mount snares examined' and then they were not
inspected again until 14 July. By this time the thylacine
population had decreased to such an extent that the tiger man
probably was spending much of his time on herding and cattle droving and general farm duties.

The dogs involved in the predation at both Woolnorth and Surrey Hills were of the large cattle dog type and had either gone wild or were part of the establishment of Aboriginal or vagabond bands—the latter probably were escaped convicts living by their wits—and both groups were blamed for stealing sheep at the Company's properties.

The tiger man would have snared dogs as well as thylacines in his work and was really what we would call today a vermin control officer. The name 'tiger man' may well have developed as a nickname but it became known throughout the state, and with the decline of the thylacine the name acquired a certain mystique which still exists today.

|

Surrey Hills

The Surrey Hills station was situated about 45 km inland from Burnie near the head of the Leven River and extended as far south as the head of the Hatfield River, to the east to Mt Tor, and west almost to Waratah. The selection of the land was based on a glowing report of the Company's first surveyor, Henry Hellyer, who spoke of good open grazing land around Valentine's Peak. The choice was a disaster and as early as 1829 Goldie, the Company's agriculturist speaking of Surrey Hills, reported 'I cannot say I have seen any country in a fit state to receive sheep ... indeed I question if some of the plains would keep stock even for a short while in summer, certainly never in winter.' The Surrey Hills overseer, Robson, recommended abandoning the property in 1832 as even bullocks died during the winter (Meston 1958; Binks 1980).

The station was accessible by a rough road which was often deep in mud, and it presented difficulties in supply as well as in clearing the thick forests, in fencing, and in stock management. Eventually the property passed out of the hands of the Company and all that remains are some typically English topographical names and a network of old roads leading to nowhere, as shown on the Tasmaps Sophia sheet. Most of Surrey Hills is now a forestry concession area. The sheep at Surrey Hills were not only subjected to predation but the climate also took its toll.

TABLE 6.6 Sheep losses from predation by thylacines and/or dogs, Surrey Hills, 1832–52

	Thylacines	Dogs	Dogs or thylacines	Total	Remarks
1832	18	45	26	89	
1833	2	172	220	394	Maximum killed in Jan.–April
1834	28	11	166	205	Maximum killed in February
1835	0	1	0	1	Only 50 sheep on property
1836	1	0	0	1	As above
1837	1	0	0	1	200 sheep on property
1838	1	0	0	1	4 sheep taken by eagles
1839	4	0	2	6	5 sheep taken by natives
1840	7	5	1	13	
1841	0	5	0	5	3 sheep taken by eagles
1842	0	0	0	0	1 lamb killed by a Tas. devil
1843	18	39	0	57	1 wether killed by an eagle
1844	4	13	43	60	1 pig killed by thylacine
1845	37	0	0	37	
1846	25	8	0	33	
1847	0	0	0	0	Losses negligible, except by stealing
1848	0	0	0	0	
1849	0	0	0	0	
1850–52	0	0	0	0	No losses
Total	146	299	458	903	

The station diaries from the Surrey Hills property of the VDL Company enable us to calculate that during 1832–49 a total of 904 sheep were killed by predators, 147 of which were blamed on thylacines and 299 were attributed to dogs, and the remaining 458 were killed by dogs or thylacines (Table 6.6). Very few sheep were killed by Tasmanian devils, the death of only one lamb being attributed to this cause (Guiler 1961a).

The number of sheep killed increased in 1833–34 and it is clear that dogs were a more serious predator than thylacines.

The percentage of sheep killed at Surrey Hills was much higher than at Woolnorth. The sheep losses in 1833 totalled 394 and in the following year another 206 sheep were lost. The sheep population in January 1834 was 1800 and we can assume that 10–20 per cent were predated annually. These percentages of loss were certainly high enough to justify some sort of control measures. The Company responded by yarding its sheep every night but eventually was forced to remove sheep from the property. In July 1834, 500 sheep were taken off Surrey Hills and the mortality rate dropped dramatically. The removal policy continued until there were only 50 sheep left by October, the casualty rate remained low and the few kills were attributed to dogs. The number of sheep was increased by 200 in February 1835 and further increased in 1839–40 when predation was kept to a minimum by yarding the stock every night. There was a sharp increase in thylacine predation in 1845–46 but this dropped again and no further losses were experienced until the records end in 1852. Perhaps the most exciting night of all occurred in 1844 when a thylacine killed a pig—such a commotion that must have been.

There was more justification for the employment of a tiger man at Surrey Hills and Hampshire than at Woolnorth, but I can find no evidence in the diaries or other sources that anything was done other than to adopt the not very positive policy of removing the stock. There is no suggestion that the tiger man was ever loaned from Woolnorth, nor is there any mention of snaring taking place or of thylacines being caught in the area.

Other properties

The station diaries and other documents from the Circular Head property do not mention thylacines for the period January 1832 to February 1851. Circular Head was a very large property which, as early as 1832, had five officers, seventeen mechanics and clerks, two female servants, forty-five assigned servants, and fifty-eight wives and children, and reached its maximum about 1893 when there were five officers, fifteen servants, and fifty-nine assigned servants plus women and children. In view of the large number of people working on the property and the absence of any reports of sightings or predation, we can assume that thylacines did not present any problems at Circular Head.

The other major holding at Middlesex Plains may well have suffered predation at the level seen on Surrey Hills as it was situated in the open-forest mountain country favoured by thylacines. However there are no records of any type from this station.

The present situation on Woolnorth

It is logical to assume that an area once inhabited by thylacines would still be attractive to them. I believe that a few thylacines may still be on Woolnorth and some accounts of our searches there will be given in Chapter 8.

Our visits to the property were not the result of inspired guesses but were the direct result of my hearing of a number of sightings during the time I was living on the property examining the station diaries. The first report was when Mr Pat Busby, who was the station manager at the time, drew my attention to the possibility of the existence of a thylacine when he, together with a station hand, found footprints on Three Sticks run on 4 June 1959. This was followed by a number of reports of a thylacine being seen on the Welcome Heath at or near the Welcome River as well as at the forest during the 1959–64 period. These names are familiar to us from earlier in this chapter. One such report is reproduced verbatim and is typical of the others.

Whilst travelling towards Woolnorth on 14 February last, about a mile past the Welcome River, an animal with very distinct stripes which ran across its back and tapered down the sides, and with a thick tail tapering down at the end as if it were dragging on the road, hopped like a dog with sore feet across the road. In the glare of the headlights it was hard to see the size of the stripes across its back, but they looked to be about an inch wide, but they could be picked out very clearly.

As I had never seen an animal like this before, I related the story to Don MacLeod, and after checking from a book, we found it to be a Tasmanian tiger.

At the time of crossing the path of our car, it was no more than five yards from us, and then it disappeared into the bush on the other side of the road. (Signed) Colin Mathieson, 1 March 1960

Inspector George Hanlon and the author looking for spoor near a fence where a sighting had been reported, Woolnorth, August 1958.

There was a further report from Don MacLeod, who succeeded Pat Busby as station manager at Woolnorth, who saw what he believed to be a thylacine about three years later at the Welcome Forest at the Old Road gates. Two further sightings were reported by station employees and these were both from the Welcome area.

These reports coming from the same district over a period of time seemed significant when considered in relation to previous accounts of thylacines staying in an area at Woolnorth, and I spent some time on the property searching for footprints and other evidence of thylacines, and we did, in fact, find spoor near Cape Grim as well as Studland Bay and Three Sticks. These findings led to the full-scale searches which will be described in Chapter 8. I am certain that there were thylacines on Woolnorth as late as 1962 and I see no reason why they would not still be there. The Company does not permit any shooting or invasion of their land and thylacines should have every chance of thriving in this congenial habitat.

The attitude of the Company towards thylacine searches

has changed drastically and we no longer have entry to Van Diemen's Land Company property. The reason given is that the property might come under some control from the National Parks and Wildlife Service should a thylacine be found there and the Company's interests would no longer be paramount. This is an understandable point of view and it is difficult to criticize the Company for the embargo.

Nowadays visitors are only allowed on the property if they join a summertime guided bus tour to a 'dude ranch' type of set-up and entry is prohibited to all other persons. Nevertheless we still hear of sightings from time to time and it is reasonable to hope that thylacines may once again be found in this area where once there were so numerous, especially as the habitat is much the same as it was at the turn of the century.

7

Some tiger tales

This chapter places on record some tales I have come across over the years, some of which have been recounted to me by men who actually caught thylacines.

In this connection we must remember that the last year of thylacine abundance was 1908 and that is now over three-quarters of a century ago. All the men who caught thylacines for bounty and knew of their habits are dead now and the tales we hear today are second-hand at best. The few captures made in the 1910–1930 period are still vivid in the memories of the participants and, for example, Wilf Batty remembers every detail of the incident when he shot the thylacine at Mawbanna. However, the stories I shall recount are not from that period but were gleaned from old-timers around twenty-five years ago when I began this research.

Many of these old men were active both mentally and physically and recalled incidents from their earlier days in the bush. Nearly all of them had figured in the Lands Department bounty accounts. The majority of them had paid the toll of a rough life living in wet conditions and suffered from rheumatism, arthritis, and other ailments, but their eyes

How many dead thylacines were brought to this house? The Pearce home, Clarence River near Derwent Bridge, February 1984.

gleamed when they spoke of the 'hard but good old days in the bush', and almost to a man would have come to help us in our search had it not been for the stiffness in their joints.

The central highlands today are readily accessible by car but this area was not opened up until fifty years ago. The Lyell Highway to the west coast did not exist until 1923, and before that date there were only a few rough cart tracks into much of the area and the region was isolated, even from the small town of Ouse. Living in this solitude in the Black Bobs–Strickland–Derwent Bridge district for many years had made the Pearce family into a reticent self-contained unit. They were not given to talking to strangers and officials from the outside world and I regard myself as privileged to have been able to talk freely to two members of the family about their experiences and about thylacines. One of H. Pearce's first questions was 'Why did you want to go and protect them bloody useless things?' and I guess that he never liked thylacines although he made quite a lot of money from them, claiming bounty on fifty-three while his relatives claimed on another twenty-three. His view was much the same as that universally adopted in various degrees of intensity by the old-timers, particularly those who were shepherds or owned sheep, that thylacines should be eliminated because they killed sheep and lambs and were of no use in the bush. There was no argument about it and they had no regrets about their deeds, nor is there any reason why they should have had, for they knew nothing of the value of a predator in a natural ecosystem and the present outlook came too late to reach them and save the thylacine, but a few did express sorrow that thylacines had vanished from the bush scene.

Around about 1953 in talking to H. Pearce he told me 'I put up a slut and three pups out of a patch of man ferns about five years ago', and I do not doubt his statement. The area where the incident occurred is now at the bottom of Lake King William and he said that he turned his dogs on to them but continually dodged the issue as to whether the thylacines were killed or not. I strongly suspect that they were, and he knew it was naughty of him to kill them.

Mr Allen Briggs of Safety Cove kindly sent me details of an interview he had with Mr J.M. Dunbabin of Dunalley on 8 September 1961 reminiscing on thylacines thus:

Tigers were a plague in the Cockle Bay area around 1870–1890 when they killed sheep. Lots of hunting and snaring them. Poisoning was no good as they never came back to a kill a second time. We caught about twenty all told but got no relief till we bought land on Forestier Peninsula and moved all the sheep. They apparently lived in the Sandspits–Pony Bottom area. French's at Buckland got about seventy. No sheep were killed after about 1890–95 on our place. The usual way was to set snares in the fences. At Cockle Bay they had a yard with a decoy sheep in it where they expected the tigers to come. They usually started to chase a sheep on bedding hill and killed when they reached the level. Not having speed he is supposed to wear his victim out with a steady chase.

The numbers he quotes are disturbingly high. The Dunabins and the Frenches presented only four animals in all for bounty and even allowing for some exaggeration, this is still a long way short of the totals mentioned above. Probably the farmers were only too pleased to get rid of the pests and did not bother too much with the bounty, or alternatively the skins could have been offered through the trade for manufacture into waistcoats as mentioned by Crowther (1883) and Laird (1968). Discrepancies appeared elsewhere in a number of interviews, and I strongly suspect that many more thylacines were killed than appear in the bounty accounts.

I have no doubt that the thylacines lived in the Sandspits–Pony Bottom area as most of the country was undeveloped at that time. Nowadays areas of it are being logged for woodchips and the habitat has been ruined. Mr Dunbabin himself told me that the Ragged Tier near Dunalley produced a few thylacines.

Mr Dunbabin also said 'Tigers were the most cowardly animal in the bush and a strand of packing twine would hold him' and once again we have a report of an animal which gives up easily and is quiet and easy to handle once caught. Also his remark that thylacines did not return to a kill was quite in agreement with the opinions of other old-timers. Setting 'necker' snares or springer snares was the usual method of capture and the tiger man on Woolnorth used to remove a paling from the fences for this purpose.

R. Stevenson, who snared on 'Aplice' near Blessington, laboriously built about 2 miles (about 3 km) of wire netting fence to guide thylacines into pitfall traps which he had dug about 2 metres deep with swinging lids. The lids tipped the thylacine into the hole. He allegedly caught sixty in this way between 1890 and 1906 (S.J. Smith 1980, quoting Stevenson). This method used by Stevenson is interesting because it shows that it was well worth while for a trapper to exert some considerable effort to catch thylacines. Although once again we have a discrepancy between the stated catch and the bounty records, the latter showing that only twenty-six were claimed and five of these were on 2 August 1894, obviously a bumper night.

The statement by Dunbabin that sheep were started on the bedding hill is interesting. Willett (1927), who trapped on The Island, also put this view when he stated that sheep were attacked whilst sleeping at night. The slow persistent chase is recorded frequently.

The Meredith family came to Tasmania in 1821 on the ship *Emerald* and immediately took up land at Cambria near

Swansea. Mrs Louisa Ann Meredith arrived in Tasmania in October 1840 and was a very active woman, interested in her new country and its natural history. She collected and painted seaweeds and plants and wrote of her experiences in Tasmania. Consequently her observations on thylacines can be regarded as perceptive and reliable. It was she who sent the thylacine to the Wilmot Zoo in Government House. This was the first live animal exhibited and was a two-thirds grown cub whose mother had been killed by a shepherd.

No care or kindness will civilise it even when taken young. Not heard of very often now. A skin was 4' 6" from head to tip of tail. Enduring but not a swift runner proceeding at a canter. A party of bushmen saw a kangaroo hopping along followed about ten minutes later by a female tiger scenting along the track and a quarter of an hour later two cubs came along the same track. (Abbreviated from L.A. Meredith 1881)

Mrs Meredith's comment that it was not possible to civilize thylacines presumably means that it was not possible to turn them into interesting or acceptable pets. This, I am sure, is very true and there are no records of anyone having a pet thylacine. When in captivity we read of them being kept in cages, sheds, and what not else, but always well locked up and never around and about the house. Most marsupials make poor household pets as they are difficult to train, some of them smell more than a bit, and the larger ones tend to take up a lot of space and are not noted for their endearing intelligent behaviour.

The description of the hunting procedure is very interesting as it suggests that the sense of smell is important, perhaps more so than is realized by those who have said that thylacines hunted mainly by sight. Also the use of olfaction by an animal would make night hunting much more effective than by using sight only. Even the cubs were using it in following the mother. Clearly they were too large for the pouch and had to tag along as best they could while she hunted. This supports the view of H. Pearce that they do not have a lair and the young follow the mother. The habit of the young following the mother may well be at a time when predation could take place, such as when the young are small and would be liable to be killed by devils.

The following description of the wreck of the *Acacia* at Mainwaring Inlet on the south-west coast appeared in the *Hobart Mercury* on 23 May 1905·

The bodies had been gnawed by Tasmanian tigers, one of which was found dead near the corpses, apparently having been bitten by snakes many of which were found near the bodies and killed by the party.

Thylacines have never been accused of necrophagy and it is likely that it was the work of devils as they will eat anything. Presumably it was a thylacine corpse which was lying near-by. Thylacines have been seen from time to time in the Mainwaring River area.

Mr George Johns of Deloraine, although too young to be on the thylacine bounty lists, spent his lifetime snaring in the Western Tiers and Arthur's Lake areas as his father had done, and he told me:

Tigers won't go over anything when they can go under it. You can set deadhead snares and they won't even strangle themselves.

The first observation is very interesting and should be noted when looking for footprints. The second relates to the docility of thylacines in snares. There are stories of wide areas around a snare being damaged and the animal having escaped but more probably this would be due to devils as they certainly do tear up everything within reach. We also have the tale about a thylacine giving a big leap and breaking the snare.

Mr S. Woods of Ringarooma wrote to me telling me of this incident which occurred in the foothills of Mount Victoria:

In 1934–35 when I was about 17 years of age a tiger went round my snares eating the possums. One night when I had a number of possum on my back a tiger followed me. I could see his eyes in the light of the torch but he followed me all the way home. I was scared. (Amended and shortened from S. Woods 1980)

Woods had seen this animal a few days previously and was able to give a good description of it. There would seem to be no doubt that it was a thylacine. The habit of following humans has already been noted, perhaps this animal was doing so in the hope of getting a possum dinner.

Another bounty claimant, Mr W. Sawford of Parattah, who died on 7 August 1977 at the age of 99 years and probably was the last tiger man in the midlands, collecting on nine thylacines, told me when I interviewed him in 1960:

> There weren't a lot about but a few could be a terrible nuisance to sheep, scaring more than they killed and scattering the mobs. I caught most of mine in the Swanston area towards the coast.

The literature states clearly that there never were many thylacines and it is good to find this supported by a bushman. His comment that more sheep were scared than were killed is supported by the statements of Cotton, Shaw, and others, who contended that good shepherding reduced thylacine depredations to a minimum.

Mr C. Thomas of Oatlands has an old diary from the 1868–1870 period and there is no mention of 'tigers' in it. Mr Thomas told me that he caught most of his thylacines east of Oatlands between Lemont and Tooms Lake and in the Ross area. The last one he killed was at St Peter's Pass. The trapping effort by Thomas, Willett, and others was mainly to the east of the midlands. Mr D.D. Davis of Ross remembered that a few thylacines used to live on Badajoz Tier near Lake Leake but he could not recall any being seen there since 1900. Here again, we find the emphasis to be on hunting in the Eastern Tiers.

Snarers worked in the Interlaken region, the bounty records show twelve thylacines being taken there. A little south near Bothwell at The Den there was a hut which had a number of thylacine skulls nailed on the wall. The hut collapsed many years ago and the skulls went too, otherwise I would have collected them.

The following comments were made to me by three east coast pastoralists:

> Kelvedon losses in the 1880s were about 200 sheep per annum, mostly stolen but about 12 per annum to tigers, mainly south of Swansea and inland towards Lake Tooms. Not many were lost on Mayfield. (T. Cotton, Kelvedon)

Lisdillon and Cranbrook had about 2000 sheep but about 10–12 per annum were mauled. Not many tigers, only about six a year were claimed as a bounty but we didn't use the back country on account of tigers, but didn't lose many sheep on account of good shepherds. (F. Shaw, Red Banks, corroborated in general outline by P. Mitchellmore, Little Swanport)

Lost 150–200 sheep per annum in the 1880s but all were stolen. No thylacines about. (R. Amos, Cranbrook)

So much for the arguments of the pro-bounty lobbyists in parliament. The three gentlemen quoted above farmed in an area from which came many bounty claims, Bicheno-Cranbrook yielding ninety-one animals, yet sheep losses were comparatively light due to good care of the flocks. These farmers suffered more losses from thieving than from thylacines predation, as did the Van Diemen's Land Company. Shaw's claim of six thylacines per annum presented for bounty does seem excessively high.

C.E. Lord, who was the director of the Tasmanian Museum, wrote in 1928:

Tigers cannot be held safely by the tail.

Other writers, for example J.S. (1862), as well as many of the old-timers, all state the opposite. It is possible to immobilize many species by holding them by the tail, and the rigidity of the thylacine's tail would enable this to be done effectively and with a certain amount of security to the holder. Try it out on a dog to see how effective it is.

Sharland writing in 1939 quoted a trapper as saying that thylacines were unable to swim. It is highly unlikely that this would be so as swimming is a facility possessed by most mammals. In fact, the Tasmanian Aborigines described the thylacine as a very strong swimmer, swimming like a dog with only the top of the head and ears above water and using the tail as a rudder (Milligan 1853). The tail is slightly laterally compressed and this led Swainson (1846) erroneously to suggest that the animal was aquatic and piscivorous.

'Thylacines eat only freshly killed food' appears time and time again in the literature and in the wild state this is undoubtedly correct. Trappers agree on this point, indeed it is one of the few on which there is total agreement. In captivity thylacines were known to eat all sorts of dead material, and this might be interpreted as indicating that they would do so in the wild but there is no evidence of any such habit.

An unusual comment was made by Nicholls (1960) who wrote that thylacines raided garbage dumps especially seeking out fat and also bread. This view has not been supported by others, but there is no doubt that thylacines did hang around bush camps, but whether to glean scraps or just out of curiosity is not known.

The next story has been paraphrased and shortened from an account by 'Correspondent' (1924) in the *Weekly Courier:*

A bushman carrying an automatic pistol for protection against tigers encountered a large male and a young female tiger near Waratah. The female sprang at him, the pistol wouldn't fire so he hit it with some wood. The male leaped at him and he hit it with a stick and it crawled away whimpering. Later, the man saw a cub and thought this was the reason for the attack.

Here is an extraordinary tale of an unprovoked attack by no less than two thylacines on a man and is atypical of all that we read or have heard of these animals. The story does not hang together, particularly the part about the automatic pistol and having time to pick up two pieces of wood to defend himself against leaping animals. I think the correspondent may have been repeating a bar-room tale. I myself do not believe the story, but I have included it to illustrate the type of tale or legend developed around a bar or a campfire which springs up from time to time and gives the thylacine a more fearsome nature than the trappers ever found it to have.

Nevertheless, the truth of this incident appeared some sixty years later when I met the brother of the bushman concerned in the affair. He was Mr C. Penny and was clearing a block of bush close to the Arthur River about 22 km from Waratah at Penny's Flats or Powcena in 1922. He started two young thylacines from their hide in the scrub and the mother chased Mr Penny. He hit her on the head with a bit of dressed timber which had a nail protruding from it and this penetrated the back of the neck of the thylacine and so 'pithed' it. Penny's brother, who now lives in Devonport, was 13 at the time and remembers the incident well. He told me that thylacines used to follow him along the track from Waratah and he was frightened of them.

This thylacine is the one on the cover of this book.

My next tale is taken from a letter submitted from Forth and the 'J.S.' is believed to have been James 'Philosopher' Smith who lived at Forth at that time (Mead 1961). Smith explored and prospected the hinterland and eventually discovered the Mount Bischoff tin field which was the richest mine in Tasmania during its period of operation.

Two tigers seen on Black Bluff in November 1862 were attacked by a setter dog which seized the tail of a tiger and was dragged off. They fought and the dog throttled the tiger. Later on another large tiger was seen but it ignored the dog. The man tried to hit the tiger with a tomahawk but it ran away and fought with the dog, both biting each other. The man hit the tiger on the head with the tomahawk but it continued to fight the dog for two minutes. The man seized its hind legs and severed the neck with the tomahawk. It was a female with four pouch young, 21 inches high, and 3 feet, 4 inches nose to rump length, and 17 inches tail length.

Tigers never attack man but will follow them for considerable distances at night but not molesting him. They retreat to a den when attacked. Sheep dogs and tigers are equally matched but the former wins the fights but is severely injured in doing so whereas kangaroo dogs win easily. (Condensed from J.S. 1862)

This story has a ring of truth about it and sounds much more authentic than the pistol episode. It is noteworthy that the thylacines did not attempt to attack the man, not even when the unfortunate animal was being walloped with the tomahawk. This female was a large animal, being well up in the size range for the species (see Chapter 3). The statement that they will follow a man at night has been encountered before—in Woods' experience—and in daylight too according to the Luckmans, H. Reynolds, and others. Retreating to a den would be the habit of any animal not able to run away.

The exploration of south-west Tasmania was largely associated with mineral and timber exploration. All of the region of about two million acres west and south of a line from Strahan to Queenstown to Ouse was untouched by roads or even tracks and only accessible to the intrepid few who walked and cut tracks for themselves and their pack-horses. One such was T.B. Moore who explored south from Macquarie Harbour in the 1880s and he named various features, Moore's Valley was one.

A fascinating photograph was published by Binks (1980) showing the bushman T.B. Moore in a posed studio photograph with his two dogs, Spiro and Spero. The remarkable feature was that Moore was wearing a 'Davy Crockett' hat made from a thylacine skin, the former owner having been killed by Moore's dogs. Spiro was a large hound and Spero was a cocker spaniel so I suspect that Spiro did most of the killing while Spero contributed much of the barking. Binks notes that the Spiro Range and the Spero River were named after these two dogs and another dog, Wanderer, a staghound, is commemorated in the river of that name.

The thylacine killed by Wilf Batty

The last authentic killing of a thylacine in which a carcase was produced was the now well-known incident at Mawbanna when Wilf Batty shot a large male animal in 1930. The story, as related by Wilf, goes much like this.

The animal had been in the area for some time and had been seen by some farm workers in the district. They were frightened of it and wouldn't go near it. Wilf was in his house and heard a scuffle and when he went outside with his gun he saw the thylacine with its head under the wire mesh around the henhouse. The thylacine moved off and went clockwise around a nearby shed. Wilf went anticlockwise and they met at the far corner of the shed whereupon they both reversed directions and met again at the opposite diagonal corner of the shed, i.e. nearest the henhouse. At no time during these manoeuvres did the thylacine hurry or seem frightened. Wilf attempted to catch it by the tail but couldn't as he was carrying his gun in one hand. The thylacine reversed direction again and headed for the fence whereupon Wilf shot it. Both his kelpie dogs were terrified by the presence of the dead thylacine and didn't return close to the house for three days. During the time it had been in the area it was seen to jump a 4 feet 9 inch (1.45 m) wire fence, touching only the top strand with its feet.

The total lack of fear shown by the thylacine is typical and has been recorded often by others. That the animal had been around for a time supports the temporary home range theory, and this is another instance of an animal hanging around outhouses either out of curiosity or more likely hoping to get a tasty chicken dinner.

Another story written in 1910 comes from north-east Tasmania:

At that time my father was connected with mining at South Mt Cameron and a popular outing was to climb Cube Rock on the Cameron Range. We always sighted tigers or hyaenas as we commonly called them and they were much larger than reports I have read about. They were as big as an Alsatian dog and would not move out of your way. We always went aroung them and always carried a rifle, though they never attacked. (J. Wiber 1977)

Quite a credible tale although I find it strange that the thylacine did not move off the track and the party had to go round it, but this further illustrates the absence of fear of humans.

Thylacines and dogs

The relationship between dogs and thylacines is yet another area where there is contradiction in the records. The most widespread opinion is that dogs showed considerable fear in the presence of thylacines and would seek shelter and protection from their owner or run away and hide (P. Le Fevre 1953), and this view was held by most of the trappers. An incident near the Walls of Jerusalem in 1958 is typical of these stories: a cattle drover was camped in a log cabin with his three dogs . . .

After dark following a scuffle in the bush, two of the dogs came into the cabin. Next day he found the third dog dead with its heart eaten out. He took his horse and two dogs to a nearby gully and the dogs ran under the horse and shortly later a tiger appeared on a rock. (Terry 1961)

Horses also were known to be reluctant to proceed further, sensing the presence of some strange animal and laying back their ears.

H. Pearce told me that thylacines could easily be killed by

cattle dogs and kangaroo dogs but the dogs had to be trained specially to do so. He said that smaller sheep dogs would find the going hard and with the smaller breeds there would be little doubt about the result. A fight between a thylacine and a bull terrier was described by Le Souef & Burrel (1927) in which the thylacine came out victor by biting off the top of the dog's skull and leaving the brain protruding.

A story from the early days of settlement presents a different picture:

A few nights ago, a hyaena, an animal so rarely seen in the Colony was found in the sheep fold of G.W. Gunning, Esq., J.P. at Coal River. Four kangaroo dogs, which were thrown in upon him, refused to fight and he had seized a lamb, when a small brown terrier of the Scotch breed was put in and instantly seized the animal, and after a severe fight, to the astonishment of everyone present, the terrier succeeded in killing its adversary. (*Hobart Town Gazette* 1823)

This tale is of note because it suggests that even in 1823 not many thylacines had been seen in the colony and gives support to the view that they were not common at the beginning of white settlement. It also supports Pearce's view that even kangaroo dogs had to be trained to attack thylacines. All we can say about the efforts of the little terrier is that we know the Scots are a tough race and love a fight. The Gunning property now is Campania House.

Thylacines and people

The general picture of the thylacine is of a docile creature which does not attack people, even under provocation, and this view was held by those trappers I interviewed. Sharland (1939) describes thylacines as timid creatures and goes on to tell of one which was surprised inside a hut near Adamsfield but made no attempt to attack the man who entered through the doorway and was only anxious to escape. He also tells us of a man walking on a log across a river when he met a thylacine coming the other way. The thylacine gave way without trying to contest the right of way on the log.

Sharland also recounts a trapper's experience when he and a companion found a thylacine in a snare. The trapper released the thylacine which made no attempt to attack him but contented itself with glaring at the companion who by this time was up a tree.

It has been mentioned earlier how the rigidity of the thylacine's tail enabled the animal to be held effectively by the tail, and this feature was used by Mrs Allen to advantage in handling a snared animal:

> My mother ... around the snare line discovered a snared tiger ... she seized it by the tail at the moment the snare broke. She was able to avoid the snapping jaws of the enraged animal by jerking its tail in the other direction as it tried to bite until her brother laid it out with a stout waddy. (Hedley Allen 1958, describing events in the Seymour district of the east coast)

There are accounts of persons being bitten by captive thylacines, even such a celebrity as David Fleay suffered this attention in Beaumaris Zoo (Smith 1980). I have no doubt that a captive or perhaps cornered animal would bite if given the opportunity, or if its young were threatened. Mrs Roberts used to move among the thylacines in Beaumaris Zoo and was never harmed.

James Le Fevre (1953) relates his reactions when he encountered

> ... a big bull tiger that showed fight when suddenly I came upon him. This fellow had just made a kill and his face was covered in blood. When I say a big one this fellow was about the size of a full grown Alsatian dog and being alone in the bush at the time and without a knife or any other weapon of defense, I took the line of least resistance and made for my camp.

And I don't blame him!

A man was bitten on the leg when he tried to separate a dog and a thylacine which were fighting at Thompson's Marshes near Chain of Lagoons on the east coast (*Hobart Mercury* 1980, reporting 1898 incident).

George Stevenson relates that he had caught a thylacine and was carrying it home in a sack while riding on horseback. The thylacine managed to turn in the sack and obtain enough grip with its paws on the horse's rump to enable it to bite Stevenson's shoulder (Stevenson 1972).

An account in the thylacine file in the State Archives tells of a Miss Murray who was doing her washing at a west coast creek when she was attacked outside her house by, of all things, a one-eyed thylacine which bit her severely on the arm but then fled after she trod on its tail. The scars could still be seen on her arm when she was in her eighties.

Brown (1973) tells the story of a west coast resident who killed a thylacine after it chased her into her hut whereupon she closed its neck in the door and belted it to death with a poker—resourceful women, these west coasters.

A thylacine shot at Fitzgerald in 1912 had bitten its killer on the foot, but a story related by Whitley (1973) must surely be in the realm of fantasy when he tells of an attack by a thylacine which was 5 feet (1.53 m) in height.

Parker (1833) reported an incident in which a thylacine entered a house at Jerusalem (Colebrook) and attempted to grab a child by the hair. It is a pity that Mr Gunning's Scottie dog had not been around at the time as he would soon have stopped that frolic.

Thylacines may be inquisitive animals as several accounts tell of them following people for some time through the bush. My friend George Smith of Zeehan told me that in 1934 two thylacines hung around his camp at the Spero River and were there every night for a fortnight.

A remarkable story was told by Harry Walsh (1979). When he was 6 years old in 1929, he was living with his parents in

... a poverty-stricken place called Pelverata in a large barn-like house with a very wide-opening door. I was in bed in daylight when a thylacine appeared. He wasn't shy—he just walked straight up to me and poked his nose in my face and I patted his head. He just stood there sniffing the big fire that we had. Then all hell broke loose, Mum and Dad ...

The thylacine got away.

Thylacines never offered much in the way of sport, although it was a change from the daily routine on Woolnorth for everybody to go off and 'shift a tiger'. The opening in 1910 of the Whale's Head Hotel at Temma, then the seaport for the mining town of Balfour, was heralded by a newspaper advertisement including tiger shooting as one of the recreations offered to patrons. There is no evidence of any takers, and knowing the country and its weather I am not surprised. This is the only attempt to use the animal for sporting purposes which has come to my notice.

The last thylacine to be bought by the Hobart City Council Zoo was caught by A. Murray in 1925, and he must be the last man alive who has a verified live thylacine capture to his credit. He said that two other thylacines were caught about the same time in the area but died, one was a very young animal. Mr Murray was born in 1900 and was only a lad when the bounty scheme ended. Miss McNamara (1983) interviewed him and he told her that he caught the thylacine near Waratah and kept it at Tarrawe (? Parrawe) and it became very friendly. He brought it to Hobart and sold it to the Zoo for £30 (actually £25—see Chapter 4) and this was the only time he visited the city—Hobart held no attractions for him!

The following is a record of the rarity of the species as displayed in the bar by a humorous landlord:

BAY VIEW HOTEL, STRAHAN, 1975
MENU
Meal Hours   1.15 a.m. — 2.30 a.m.

Consomme of Blue Tongue	$0.50
Parrot Pie	$1.50
Kentucky Fried Ferret	$2.50
Crumbed Coot	$2.00
Sweet 'n Sour Devil	$4.50
Curried Cormorant	$0.50
Fur Burgers	$5.80
Tasty Tassie Tiger	$2420.50
Goanna Grits	$0.80

The price for a 'Tasty Tassie Tiger' has skyrocketed since then.

8

Expeditions and searches

Since being granted partial protection in 1930 there have been few authentic reports of thylacines being killed or captured. The killing of a thylacine at Mawbanna has already been noted and another allegedly was captured in the Florentine Valley in 1933 and sold to the Hobart Zoo where it died in 1934. There is no record whatsoever of this transaction in the minutes of the Hobart City Council Reserves Committee and it is highly improbable that this story is correct. However, Elias Churchill caught a female and three cubs in the Florentine in 1925 and displayed them privately around the state.

There have been rumours of thylacines which were caught but not brought to the notice of the authorities or museums, one such being a young animal which was snared on the Western Tiers near Cressy in the mid-1960s and brought into Launceston but subsequently released. Other captures were reported by J. LeFevre (1953) from Blue Tiers where he caught one in a snare, and G. Johns (pers. comm.) remembered two being snared in 1960 and 1961 at Mersey Lea and Golden Valley respectively. One also was snared at Milabena in 1976.

These unsubstantiated rumours of capture occur from time to time and they are not reported because of fear of prosecution under the protection laws. Such legislation may serve to protect the target species but it also hinders the collection of useful information. The specimen brought into Launceston was a case in point. The then director of the Launceston Museum heard of it and tried his hardest to find out more about the incident as did Inspector N.D. McIntyre but neither of them could locate either the animal or the person responsible.

Since the days when thylacines became rare, sighting reports have been received by government departments, museums, the University of Tasmania, the press, and various magazines. The number of these reports has increased in recent years to almost a flow.

The reports have been made by persons who really believed that they had seen a thylacine, but not all stand up to further detailed examination, and a few are hoaxes. I am inclined to believe those reports which emanate from the same area over a restricted period of time from people who are totally unknown to one another and who could be considered to be reliable witnesses. For example, during the early 1960s no less than four reports were received from the Collingwood River–Cardigan Flats area from a truck driver, a surveyor, a tourist, and the Inspector of Police stationed in Queenstown.

A detailed analysis of all sightings was carried out by S.J. Smith (1980) who devised a points system for reports and gave a rating of good or otherwise, and he awarded a good rating to 106 sightings out of a total of 320 during the period 1934–80. The number of sightings recorded by S.J. Smith has been added to the number of alleged sightings in my files (Table 8.1) and these are shown in Fig. 8.1.

The distribution of sightings in relation to time is noteworthy, particularly the peak in 1970 when thirty-two were recorded. The number of sightings increased only slowly from 1934 until 1960 when it rose rapidly to the 1970 peak, then a decline set in followed by another peak in 1980. The very high peak in 1970 was due to the efforts of the Griffiths–Malley search and their publicity requesting this type of information. I suggest that the increases in 1960 and in 1980 were due also to public awareness of the expeditions at those times.

TABLE 8.1　Alleged thylacine sightings, 1940–82

Date	Locality	Report	Source	Rating
1940	Frodsham's Pass	Vocalizations	C.B. Stacey	x
27 May 1951	Great Lake at the Pencil Pines	Sighting	R. Milburn	
? 1951	Lyell Highway near Princess River	Sighting	R. Targett	x
4 Aug. 1952	Blackwell's Logging Camp, Oonah	Sighting	?	
20 Aug. 1952	House's run, 25 km south of Milabena	Thylacine eating a calf[a]	?	xxx
1952	Jerusalem Walls	Sighting	R.R. Dixon	
9 Aug. 1953	Blue Tier	Caught in snare	J. LeFevre (1953)	x
Sept. 1953	Western Tiers, 6 km from Auburn	Sighting	L.O. Keefe	xx
Nov. 1955	Loongana	Sighting of female + 2 cubs	A.N. Chaplin	
11 Jan. 1956	Traveller's Range[b]	Sighting	Dr Anderson and party of 3	xxx
11 Feb. 1956	11 mile peg from Queenstown on Strahan Rd	Sighting	Inspector Kearney	xx
Apr. 1956	Que River	Sighting	D. Nicholas	
1 Jun. 1956	Hibbs Beach[c]	Footprints	*Hobart Mercury*, 7 Jan. 1957	x
Mar. 1957	Cardigan River	Sighting		xx
1958	Jerusalem Walls[d]	Sighting	J. Johns	xxx
Mar. 1958	Cardigan River	Footprint	L. Stephens (Ewence 1961)	xxx
18 May 1959	5 km south of Trowutta	Sighting	Devitt & Malley (*Mercury* 19 May 1959)	
29 May 1959	Mawbanna Rly Stn	Sighting	P. Ralston	
Oct. 1959	Zeehan–Queenstown Rd	Sighting	A.R. Fairchild	
1959	Travellers Range	Footprints	T. Pearse	xxx
Dec. 1959	Coles Bay Rd	Sighting[e]	A. Machin (Ewence 1961)	xx
1960	Mersey Lea	Snared	G. Johns (pers. comm.)	xx
4 Mar. 1960	9 mile peg on Queenstown–Strahan Rd	Sighting	Ewence 1961	
1960	Near Manuka River[f]	Sighting		xxx
1961	Golden Valley	Snared	G. Johns (pers. comm.)	xx
1961	Legerwood Road	Sighting		
Dec. 1962	Woolnorth[g]	Sighting	B. Ritchie	xxx
1964	Collingwood River	Sighting	G. Fisher	xx
Mar. 1966	Henty River	Footprints	L. Larcomb, K. Rogers	xx
1968	Fingal	Sightings	N. Sutton	xx
16 May 1969	Lewis River	Sightings		xx
12 Jun. 1969	Woolnorth	Footprints	K. Harmon	xxx
1969	Piper's River	Sighting	Mrs Targett	
1969	Fingal	Sightings	N. Sutton	xx
7 Jan. 1970	North of Pt Hibbs	Sighting	D. Whayman	
1970	Rapid River near Gibson's Plains	Sighting	W. Sawley	xxx
1970	As above	Sighting	R. Reid, L. Farrelly	xxx
1970	Welcome Swamp	Sighting	VDL Co. employees	xx
1970	Cluny	Footprints	Maj. Bowden	
Apr. 1971	Liena	Sighting	G. Mackey	
July 1971	Copley Dam	Sighting	G. Miller	
Jan. 1972	Mt Roland	Sighting	D. Mackowski	
1972	Dunalley	Sighting	Mrs Jarrott	
Feb–Mar. 1973	Flinders Creek	Sighting	M. Marsden	

TABLE 8.1 (cont.)

Date	Locality	Report	Source	Rating
1974	Lake St Clair	Sighting	W.G. Sow	
1974	Lisdillon	Sighting	A. Thompson	
12 Jan. 1975	87 km post near Orford[h]	Sighting	D. Hallam	xxx
Mar. 1975	Milabena[i]	Sighting	B. Casey	xx
11 Oct. 1975	Sister's Hills	Sighting	M. Molesworth	xx
16 Nov. 1975	Sister's Hills	Sighting	M. Molesworth	
1975	Mt Lloyd	Sighting	E. McDiarmid	x
Mar. 1976	Milabena	Snared	Mr Poke	
4 Jun. 1976	South of Pieman River	Wallaby kill	J. Barrett	xxx
1976	87 km post near Orford	Sighting	G.B. Harrison	xxx
Feb. 1978	Waratah Highway	Sighting	Larkin 1978	
Nov. 1978	Hogg's Is., west coast	Footprints	? Walker	
Feb. 1980	Dawson's Siding	Sightings	N. Forster	
Feb. 1980	Mulcahy Beach	Sighting	? Walker	
1980	Lake Chisholm	Footprint	D. Robinson	
Mar. 1981	Mt Eliza	Tracks	P. Salter	
Mar. 1981	Mainwaring River	Sighting	J. Yates	
Nov. 1981	Lyell Highway	Sighting	W. Hales	
Mar. 1982	Avoca	Sighting	T. Freeman	
Oct. 1982	Emu River	Sighting	N. Clifford	

Notes: Each sighting has been given a rating: x–person known to have a knowledge of the bush and to be a reliable witness; xx–interviewed by police or officers and found to be reliable; xxx–either of first two categories supported by later sightings or field evidence.

[a] This is the only recorded incident of a thylacine eating and, presumably, killing a calf.
[b] See 1959 record.
[c] See 1970 record.
[d] See 1952 record.
[e] Sighting confirmed later by another viewer, Mr McRae.
[f] I visited this area about a month later, and heard the yapping hunting noise made by thylacines.
[g] This sighting was made at a place where several sightings have occured over recent years
[h] See 1976 sighting.
[i] See the 1952 incident.

The geographical distribution of sightings shows that most of them occurred in the northern part of Tasmania with some emphasis on the areas of former abundance in the north east, the central highlands, and the north west. There were sightings from the southern part of the east coast. It is noteworthy that sightings were generally distributed over all of the state just as were thylacines in their days of abundance. The few sightings from south-western Tasmania were made on the coast, which is where they were found last century.

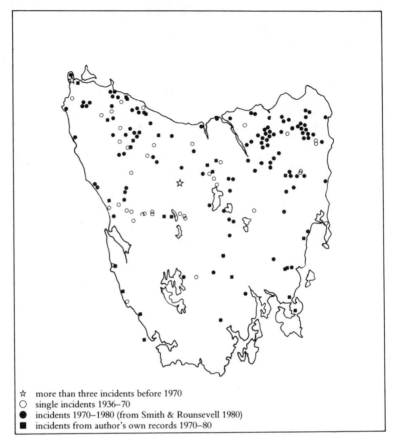

☆   more than three incidents before 1970
○   single incidents 1936–70
●   incidents 1970–1980 (from Smith & Rounsevell 1980)
■   incidents from author's own records 1970–80

Fig. 8.1   Outline map of Tasmania showing the location of alleged thylacine sightings, 1936–80. Note the concentration of sightings in areas of
former abundance.

Most of the sightings as recorded by S.J. Smith occurred in
areas of low-lying ground up to 200 metres in height (145 out
of 320 incidents) with only 43 occurring above 600 metres.
This should not be interpreted as indicating an altitudinal
preference by the thylacines but it is more likely to be due to
a combination of the activities of humans and the distribution
of roads, both being largely in the valleys. In fact, we know
that the species was distributed throughout both low and

high country, although most of the carcases presented for bounty payments were caught on the central plateau or in the foothills of the Ben Lomond massif.

S.J. Smith found that the sightings occurred at all seasons of the year, being equally distributed throughout the year, and were made by persons in a wide variety of occupations. Most of the sightings were made by residents of the area in which the incidents occurred (73 per cent), and of these sightings 56 per cent were made by persons who were alone on the occasion, while 44 per cent were by groups of two or more persons. On one occasion near Trowutta two carloads of people saw the thylacine at the same time.

S.J. Smith's analysis showed that the length of the observation was very brief, about 3–5 seconds being the usual time. Many of the sightings took place at night which meant that the animal was close to its viewers. The brief nature of the observation combined with the limited field of view of car headlights and the necessary concentration on the road make the identification difficult and sometimes dubious. I remember seeing a wallaby on the Corinna road one night and all of our party were sure that it had stripes across its rump but striped wallabies are unknown. All I am suggesting is that nocturnal observations must have some doubt attached to them, just from the very method and difficulties of observation. I would give a high rating to about 10 per cent of the reports which I have received, lower than that given by S.J. Smith on sighting reports.

Some reports which are widely believed to be true at the time turn out to be false and I have not included these in Table 8.1. For instance, a helicopter crew of two saw an animal on a beach in the south west on 6 January 1957, and photographed it and this was published in *Pix* of 16 February 1957. I was not convinced that the photograph was of a thylacine as the animal had a bushy tail and appeared to be too dark in colour. Subsequently I interviewed the helicopter pilot, Captain Holyman, who would not state that he was certain of its identity. Later I heard of a fisherman who had lost his dog in that vicinity at the end of December when they had gone ashore for exercise and the dog failed to return, this was just a couple of weeks before the photograph was taken.

At one stage I would dash off into the bush to seek evidence at the site of every report, but so little was found that this became a most time-consuming business yielding no results. The same exercise was carried out by S.J. Smith during his investigations but he too found them equally unrewarding and time consuming. Nowadays I prefer to wait until there are several reports from an area and then go looking for field signs.

The status of the thylacine must always remain in doubt as long as there is a constant dribble of alleged sightings. Both S.J. Smith and the Griffiths–Brown combination concluded that the thylacine was extinct because they did not find any positive evidence of its existence, but I do not accept their conclusion.

Expeditions, searches, and investigations

The quest for the thylacine has taken three forms based on the level of activity and energy expended, and it is appropriate to classify these forms as searches, investigations, and expeditions. The last are on a large scale which may last for some time, often for months, and usually with some sort of official blessing, and may incorporate several searches and investigations. The investigation lasts for a short time and often takes the form of an inquiry into a sighting report. A search is a small-scale expedition lasting only a matter of a few weeks and is not necessarily confined even to an alleged sighting.

Recognizing the above divisions we can then set out a chronological table of events, each of which will be examined in more or less detail (Table 8.2).

The dominant consideration in any search or expedition must be for the safety and security of any thylacine which may be found. This is a point which often tends to be overlooked or ignored by enthusiastic searchers but it is a topic which causes much concern in the mind of any faunal administrator.

The interest in the thylacine is worldwide, as is illustrated by the number of newspaper news items and articles which appear from time to time. I list small selection just to make

TABLE 8.2   Major thylacine searches, investigations, and expeditions, 1937–1980

	Leader		Status
1937	Summers	Search in north west	Official
1937	Fleming	Search in west coast ranges	Official
1938	Fleming	Expedition to west coast	Official
1945	Fleay	Expedition to west	Private
1957	Guiler	Search in Derwent Valley	Official
1958	Guiler	Search in Rossarden area	Official
1959	Guiler	Investigation at Trowutta	Official
1960	Guiler	Investigation at Woolnorth	Official
1961	Guiler	Investigation at west coast	Official
1963	Guiler	Investigation on West Coast Highway	Official
1963	Guiler	Expedition to west coast	Official
1966	Guiler	Second expedition to west coast	Official
1966	Guiler	Search at Whyte River	Official
1968	Griffiths and Malley	Expedition to all districts	Private
1973	Sayles	Search in north west	Private
1978	Tangey	Search in north east	Private
1980	Smith	Investigaion and search	Official
1980	Guiler	Search and expedition	Official

the point: Lukas 1963; *The Times* 1979; Wilford 1980; Anon. 1980; *Washington Post* 1980; Vechtmann 1980; *Hong Kong Standard* 1980; *Haitch* 1981; *Pretoria News* 1981. Should a thylacine be discovered there will be fantastic pressure from television groups, newspapers, magazines, entrepreneurs and prospective moneymakers, would-be hunters, and heaven only knows who else including well-meaning but equally dangerous nature lovers. In order to thwart all these people, at least temporarily, it is essential for the expedition organizers to be in a position to apply a complete news blackout on any discovery until such a time as the future security of the animal and its habitat is ensured, and disturbance in the area can be kept to a minimum.

On our researches and expeditions it always worried me what we would do if we actually caught a thylacine. If the capture had taken place on crown land then the problem would have been relatively simple as it would have been a matter for government action, but on private land it would have created very serious problems. For this reason private searches have never been encouraged and I hope they never will be.

The Summers search

The state fauna regulations used to be administered under the *Lands Act* 1909 but in 1927 the government passed the Animals and Birds Protection Act which established a Board and a field staff to police the Act. The Board always was fortunate in that it obtained experienced bushmen on second-ment from the Police Department for these duties, largely due to the interest of the Police Commissioner of the day who was always a member of the Board.

Concern was being expressed in the mid-1930s about the rarity of thylacines and the Board despatched Sergeant Summers, together with another policeman and a bushman, to the mountainous country inland from the north-west coast. They searched Middlesex Plains, the old Van Diemen's Land Company property, then moved further west to the country between the Arthur and Pieman Rivers, working as far inland as Waratah. They reported seeing signs of thylacines, particularly in the Arthur–Pieman area inland as far as the Lofty Ranges and the Donaldson River. They received several sighting reports from the same area but did not see any thylacines themselves. They recommended a sanctuary in the area but this was not followed up and there still is no game reserve of any status whatsoever in the region.

The 1937 Fleming search

The Fauna Board, no doubt encouraged by the results of the Summers efforts, sent Trooper Fleming off with a prospector, L. Williams, to search the west coast ranges in the area of the Raglan Range–Frenchman's Cap. They found tracks in different places but no thylacine was sighted. A spoor found indicated that of a small animal so they concluded that a breeding population existed and Fleming believed that there were four animals in the area (Fleming 1939).

Both this search and that carried out by Summers were in very difficult country and conditions must have been trying. Trooper Fleming was a very strong man and highly capable in the bush but even he must have found the going particularly rough. Incidentally, the rank of Trooper no longer exists in the Tasmanian Police Force.

The Fleming expedition, 1938

Still feeling hopeful and spurred on by the enthusiasm of Arthur Fleming, the Fauna Board launched yet another expedition in 1938. Their efforts were on the fringes of the area searched by Fleming the previous year in the Jane River–Prince of Wales Range. The expedition of six included Michael Sharland and two goldminers from the Jane River goldfields. This expedition worked for eleven days and saw footprints in several places, even around the hut in which they were camped. They were sufficiently convinced by what they saw to recommend a sanctuary in the Raglan Ranges–Collingwood Ranges and in the Frenchman's Cap area (Sharland 1939; Fleming 1939). The Frenchman's Cap is now a national park, but not as a result of these recommendations and the park does not include the Raglans.

The Fleay expedition

David Fleay, the then director of the Sir Colin Mackenzie Sanctuary at Healesville (Victoria), set off in November 1945 in an attempt to catch a pair of thylacines. He was accompanied by the indefatigable Fleming as well as by Michael Sharland and several others, and they moved off to the Jane River which had now been abandoned by the miners. The members worked very hard lugging large traps around and these were baited with bacon, live fowls, and meat. They saw no evidence of thylacines but were amazed to find large numbers of snares set in the bush. They attributed the absence of thylacines to the poison the trappers had laid to rid their traplines of devils.

The Jane River area was abandoned and the expedition moved to the Lake St Clair region; the party trapped around the Hugel Lakes where an alleged sighting had been reported a year or two previously.

A sighting at the Collingwood Flats sent the expedition scurrying there in January 1946 and between then and March they worked their way back to the Jane River. Live decoys

were used and semicooked livers, hearts, and other 'delica-
cies' were dragged about the bush in an attempt to attract
thylacines to a variety of traps and snares. I would imagine
that cooking the lures would be highly necessary in the summer. However, cooked or not the lures failed to attract any
thylacines and one set of footprints was all that the searchers
got for their efforts, although they caught many individuals
of almost all of the other Tasmanian mammal fauna (Fleay
1946). After a few more weeks during which a wallaby was
killed right next to one of their traps, allegedly by a thylacine, a rather disconsolate group left the traps in position
where local bushmen kept them in operation for some time
further.

I pay tribute to these early expeditioners, particularly to
my friends Michael Sharland and the late Arthur Fleming.
They endured tough conditions which were utterly miserable
at times and this was without the benefit of modern light-
weight gear and waterproofs. Arthur Fleming humped a 70
pound (approx. 32 kg) pack on these trips over rugged country which is difficult even without a pack—a feat which is
still talked about around campfires. These men had to walk
everywhere, whereas nowadays we can drive a four-wheel
drive vehicle nearly to the Jane River. These factors make
their feats even more remarkable.

The Derwent Valley search

On 30 September 1957 two sheep were killed in rather
mysterious circumstances on Mr J. Walker's property at
Tanina, near Broadmarsh. Two other sheep had been killed
about one month previously during a period of snowy
weather on the bluff behind the property and a neighbour
reported that he had lost a sheep in similar circumstances
about two years earlier.

The sheep had all been killed by having the throat eaten out
very cleanly, there was no blood on or near the carcase and it
had probably been lapped up by the killer. The nasal bones
also had been eaten away and there was no wool torn off
the victim or other signs of 'worrying' (Guiler & Meldrum
1958).

Head of a sheep killed by a thylacine, Broadmarsh, 1957.

Investigations in the area yielded some footprints which resembled those of a thylacine and some more possible ones were found further back into the Black Hills. A lamb was discovered on 17 October of the same year which had had the ribs of the left-hand side eaten away and the liver also removed. After this no further kills of this nature were reported. When later I described these kills to H. Pearce he had absolutely no hesitation in saying that they were typical of thylacine kills.

In the course of our investigations we were told that Mr L. Tringrove had seen thylacines on his Black Hills property. When interviewed he said that he had had several visits from the animal and that one night while he was sleeping in his caravan on the property he heard a rattling in the garbage can about 10 feet (3 m) away. He arose and switched on a powerful torch and plainly saw a thylacine. It was holding a loaf of bread in its mouth and was grasping it in its gape. The animal bounded about 3 metres, turned around and looked at Tringrove, then slowly moved away, being watched for 50 metres or so for a period of nearly 2 minutes. Mr Tringrove was emphatic that the animal had stripes but could not say how many. He also said that it 'turned all of a piece like a ship'.

We had a trap constructed and carted up into the Hills where it was placed near Tringrove's camp and serviced regularly from February 1958 until September 1958 when it was removed because of lack of results. Subsequently the property was sold and the trees were cleared, converting the trapping area into open pastures and considerable hunting activity has developed over the whole area.

Brown (1973) does not accept that these killings could have been those of a thylacine and attributes them to a dog, stating that a large Alsatian dog was caught in a trap at the 'site' about four months after the killing ceased. I understand that the dog belonged to Mr Tringrove, and I do not see that the event has any bearing on the killings.

Looking back at all the circumstances, I am confident that the killings were not the work of a dog. Dogs are noted as very 'messy' killers and often severely damage their prey before killing it. Farmers know and recognize dog kills and I am sure Mr Walker would have had experience of these.

This area, although close to Hobart and New Norfolk, produced thylacines in the bounty days, no fewer than eleven being paid in the Broadmarsh Valley. It is not far across country from the valley to the Mt Field National Park–Florentine Valley district and most of this distance could be travelled through light scrub and woodland.

The Rossarden investigation

The Fauna Board was notified by Inspector McIntyre on 30 September 1958 that a thylacine was suspected of being in the Rossarden region in the foothills of the Stack's Bluff–Ben Lomond massif. A group consisting of Inspector McIntyre, an assistant, and myself visited the area to investigate. Footprints had been seen and covered over with bark on a logging track outside Rossarden by a contractor, Mr J. Blacklow. There were eight prints, five of which were 'plastered'. Only one print was especially clear and was that of a front paw of a thylacine. The length between strides was 29 inches (73.7 cm), indicating a three-quarter grown animal.

Trap being carried into position, Rossarden, December 1958.

The Fauna Board gave permission for Jim Blacklow to try to catch the animal and a trap was sent to him by rail from Hobart. There is no point in trying to fool people about intentions as a railway porter when he saw the trap realized immediately what was going on and was reported to have commented 'Are you bastards going to trap a tiger?'

The trap was baited and left near a cave where the tracks indicated that a large animal had been resting, other caves nearby produced devil droppings, wombat prints, and other unidentifiable prints and scats. Jim nailed sheep heads to trees as lures and all these were eaten by some animal, including a head which had been placed about a metre up a tree. The claw marks of some large animal could be seen on this tree. I suspect that devils were the culprits.

During one of our visits we were accompanied by none other than the intrepid Arthur Fleming, still seeking the ever-elusive thylacine.

The trap remained in this vicinity for several months but, in spite of dragging bullock livers around and about, we saw no further positive evidence of a thylacine. The area was kept under general observation until about September 1959 when the trap was removed and the area was abandoned. Inspector McIntyre sighted tracks here again on 1 November 1960. On one of our visits, 5 September 1959, we found a dead wombat outside one of the caves. It had severe throat and groin injuries but was uneaten. The wombat was about three-quarters grown and was a male.

There was a sequel to these events when some ten years later Mr Noel Sutton of Fingal claimed that he had seen three young thylacines in the bush a few kilometres from Fingal, which is not far from Rossarden. Subsequent efforts by Mr Sutton and several field trips by me failed to find anything. Noel was most anxious to find positive evidence and continued his searching until ill-health forced him to quit his bush work, but found no more signs of thylacines.

The Rossarden–Fingal area was renowned for its thylacines for, after all, did not Davies of Fingal claim to lose 20 per cent of his sheep every year? Malahide estate, right beside where Noel Sutton was working, killed no less than fifty thylacines in pre-bounty times. The nature of the country around Fingal–Rossarden has not altered much in recent times and there is no reason to assume that it would not still suit thylacines, especially as there is plenty of game for their food requirements.

A sighting was reported in March 1982 near Avoca, which is some 28 kilometres from Fingal and less than 20 from the Rossarden area.

The Trowutta investigation
The Trowutta area over recent years has been the scene of a number of alleged sighting incidents and it is near here that the last known killing of a thylacine—by Wilf Batty—took place. It was therefore no surprise to hear from Inspector McIntyre that Mr A.J. Walters claimed to have seen a thylacine standing on the woodheap at Trowutta Mill. The incident occurred about the beginning of August 1959 and we visited the area on 26 August. We could find no supporting evidence either at the Mill or on a walk along tracks to the Arthur River.

Mr Walters, then about 60–65 years old, was a snarer by

Scene of Noel Sutton's alleged sighting, Fingal, February 1969.

trade and once caught a thylacine on Valentine's Peak years ago. There is no doubt that he would have been able to recognize the animal on the woodheap as being a thylacine.

Sometime later a large animal was caught in a snare set by James Malley, but escaped in roughly the same area. Inspector Hanlon and I went to investigate with James and found that the animal had torn down bushes in its efforts to free itself, but we were unable to identify the animal. Perhaps it was this incident that started James off on his hunts for a thylacine. It is unlikely that the animal in the snare was a thylacine as the torn bushes at the scene do not fit in with the agreed opinion of thylacine hunters that they just gave up in a snare. The animal was probably too large to be a devil and may well have been a dog; dogs are known to fight strongly in a snare.

About this time we carried out a number of other investigations into alleged sightings but found no supporting evidence. Sightings from such diverse places as Que River, Cuckoo, and Dunalley were checked and all must remain as question marks of varying sizes. Other incidents were followed up and could be said to be both dubious and doubtful, and are not included in Table 8.1.

The first Woolnorth expedition

We have already seen that Woolnorth was a favoured thylacine habitat and the historical researches gave us a reasonably good idea of where to start looking for thylacines on the property. In addition, we had the recent evidence of sightings supplied to us by the station staff. It would be logical to expect that thylacines still occurred there and this was especially true since many of the runs at that time were still in an undeveloped state and were more or less in the same ecological state as during the days of thylacine abundance.

The initial impetus for our Woolnorth investigation came from my visit to the property to examine the station diaries; it was during that time that Mr Pat Busby, the station manager, told me of some recent sightings which sounded so good that we decided to have a look at the area.

The expedition left Hobart early on 7 November 1959 and established camp that day on the track to the Studland Bay run. The very next day we saw tracks at the northern end of Studland Bay and set snares in likely places. These snares were set to catch the animal by a leg and were not 'neckers'.

Inspector George Hanlon setting a snare, Studland Bay, 1959.

The only catch of note during the week we were there was a wallaby which had been eaten and killed (in that order) by a tiger cat, the rectum and lower guts of the wallaby being devoured. During this trip we found a wombat with badly flyblown bites on its lower back region and put the animal out of its misery. The bites were about 3 inches (86 mm) long and we could not recognize the cause.

We did several night patrols both with and without lights, the most noteworthy being on a moonlight night down at Studland Bay when we were walking along the coastal grasslands without lights. Although conditions were ideal for feeding, the game was behaving in a peculiar fashion, all of the animals being very timid and not moving more than a couple of metres from cover while the slightest movement sent them scurrying for shelter. We never observed similar behaviour on other nights or on any future trips.

I think that on this expedition we probably got as near to a thylacine as I have ever been. One morning it rained until 6.00 AM and we went off on foot later in the morning. About 10.00 AM we passed a muddy waterhole where we found a

Coastal dunes, Studland Bay, a habitat much favoured by thylacines.

clear thylacine footprint which had no rain specks on it. This was about all we saw but it was enough to encourage us and within about four hours of our departure from the property Pat Busby found some 'interesting' footprints on the Three Sticks run (Busby pers. comm.).

The second Woolnorth expedition
A second expedition, encouraged by the results of the first as well as by Busby's footprint discovery and a further sighting report, set off in May–June 1960. This time there were three members of the group, Inspector Hanlon, Tom McMahon of the University Photographic Section, and myself.

The sighting had been at the Old Cottage down towards Cape Grim where the animal had been seen in a nearby paddock, and later seen moving along an adjacent fence line. The cottage became our base and we set snares along the fence and generally scouted around the property, particularly on the Three Sticks run.

This expedition was the first to make use of automatic cameras to try to obtain photographs. The project was experimental rather than the major part of the expedition as we wanted to find out the practical difficulties in this type of exercise. We arranged a 16 mm Bolex movie camera connected to a trigger set at a gap in the fence beside the Old Cottage. The camera was set on single frame and used black-and-white film with electronic flash. The arrangement worked well and we obtained some good shots of possums and devils.

The snares brought us better luck than previously, catching wallabies and devils. One wallaby had been eaten while in the snare and had its head, heart, lungs, and liver eaten although the remainder of the body was intact. We could not find any 'sign' to give us a clue as to the perpetrator of this deed but I suspect that it may well have been a devil.

The third Woolnorth expedition

By now we were thoroughly familiar with Woolnorth and we returned again on 17 February 1961 and this time the emphasis was on investigating rather than snaring. The encouraging results that we had obtained with the cameras led us to develop the Mark II automatic camera.

We constructed five units which consisted of a G.45 aircraft 8 mm movie camera arranged with lights covering gaps which we made in the fence on the Three Sticks run. The cameras were operated by a treadle buried on the trail under the fence. When an animal stood on the treadle it closed contacts and in so doing it operated the camera and lights.

The apparatus was not very successful as it had some defects which we could not eliminate. Firstly, the cameras operated off 24 volts while the lights, which were car headlight sealed-beam units, worked off 12 volts. We had to tap off 12 volts from the 24-volt lead, and this was done using a clip and wire resister. The wire got hot under running and we nearly caused a bushfire. We had some trouble from arcing of the contacts, resulting in their fusion, and this also was a possible fire hazard as sparks jumped during the making and breaking of the contact, especially if the pressure on the treadle was slight. The labour of carrying the two 12-volt accumulators around the bush got a bit wearisome, but we did get some reasonable pictures although the camera lenses were of not very high quality.

The advantage of the system was that it was simple and

cheap, and it was all we could afford in those days. The aircraft cameras were from RAAF disposal stores and cost only £5 each, the accumulators were the most expensive part and we managed to scrounge them on loan. One of these units has been lodged in the Queen Victoria Museum in Launceston.

We saw no signs of thylacines, although for a week we trudged over much of the property from Mt Cameron through Studland Bay and the Three Sticks to Cape Grim and the Harbour.

The fence line on which we set the cameras is of some historic interest as it was made from wooden pickets inserted between twisted wires between the posts. This may well have been a very early fence as G. Wainwright told me that snarers used to catch game by removing a picket from the fence and then setting a necker snare in the gap. Perhaps this was the very same fence line along which thylacines were caught for bounty, but we caught nothing worthy of note on our cameras.

Subsequent Woolnorth investigations

Several short trips of a few days duration each were made to Woolnorth to check up on the sporadic sighting or footprint reports that came to us, and as ever we kept an eye out for whatever might be around. Most of the sightings at this time occurred mainly on the road in the region of the Harcus River, Welcome Heath, and Welcome River. These names are familiar from the events described in Chapter 6. One report speaks most definitely of seeing a thylacine quite clearly for about 2 seconds at the side of the road (Ritchie 1961). The other sightings were equally fleeting but nevertheless the viewer always was emphatic about what had been seen.

On one trip to Woolnorth with Inspector McIntyre we thought that we really had made it at last. When walking across a paddock near the Old Cottage we spied a largish animal which looked like a thylacine, stripes and all. Fortunately Mac had a pair of field glasses with him and we were able to identify the animal as a feral cat which was quite the biggest pussy cat that I have ever seen. The locals assured us that it was enormous and that it had even attacked their dogs on one occasion.

The winters at Woolnorth tend to be severe and rather miserable with westerly winds, and the heavy rain makes the low-lying paddocks impassable for vehicles. At one stage the manager used a Bombardier to get around in the winter but the mud won in the end and horses once again came into their own. The higher and sandier parts of the property, especially along the cliffs from Cape Grim to Studland Bay, are delightful, especially in summer with the fresh wind off the sea making walking a real pleasure, and under a clear blue sky the albatrosses and gannets can be seen fishing several hundred feet below in the deep blue sea.

The Sandy Cape investigations

Two fishermen, L. Thompson and B. Morrison, were camped at Sandy Cape in August 1961 in one of two adjoining huts and were engaged in catching fish to sell to the crayfish boats for use as bait. The other hut was used to store the nets and catch. Morrison heard a noise in the middle of the night in the next hut, and went out with a weak torch and a piece of 4 × 2 wood in his hand. He saw two eyes gleaming and hit them very hard with the wood and returned to bed. Next morning he found a young half-grown male thylacine lying stretched out dead on the floor of the hut. He dragged it outside and covered it with a sheet of roofing iron which he weighed down with a baulk of timber and then went off with Thompson and reported his find to the fishing boats anchored nearby. According to Morrison, he and Thompson were plied with grog and when they returned to their camp not only was the animal missing but a pair of paddles also had been removed. Morrison suspected that somebody from the boats heard of the tale and came ashore and removed the carcase.

Thompson, who did some geological collecting for Melbourne University, scraped up some blood and hair together with a lot of sand, and about a week later these eventually came to me at the University of Tasmania. The blood was quite rotten by this time but examination of the hairs showed that they were not those of a devil, being longer and lighter in colour and the cuticular scale pattern could have been that of a thylacine.

Morrison turned up shortly afterwards in Hobart and was

subjected to police questioning on the matter but his story remained unchanged in every detail. In company with Inspector McIntyre I took him to the museum where by chance they had a live devil in a cage and on being shown the animal Morrison was emphatic that he knew what it was and that it was not the same as the animal he had killed. In fact, he was quite scathing about it all and implied very strongly that we thought he was a fool.

Immediately after this, Inspector Hanlon, R.G. Hooper, and I took off for Sandy Cape and found the scene exactly as described by Morrison, even to the positioning of the iron and the baulk of timber.

The Fauna Board was concerned that the carcase might be offered on the market and all of the Australian fauna authorities were warned of this possibility. Some information was given to the police to the effect that the carcase had been removed by a fisherman whose name was given and that this fisherman had, in fact, offered it to Sir Edward Hallstrom of the Taronga Park Trust. Sir Edward very properly refused to have anything to do with it and the carcase then allegedly was dumped at sea off the New South Wales coast, but this is hearsay.

I have given this story in full since it differs from that given by Morris (1962) who got his information second-hand and stated that the animal had been trapped and was so savage that it escaped. The matter became clouded when S.J. Smith (1980) interviewed Thompson who said that it definitely was a devil that was killed, although in 1961 he had said that he was unable to identify the animal. Morrison could not be re-interviewed as he was drowned in 1980. The change of opinion by Thompson is strange and puzzling and does nothing to clarify the situation. Smith inclines to the Thompson story and points out that there are plenty of Tasmanian devils at Granville Harbour, about 50 km south of Sandy Cape, and at the Cape. I do not accept Thompson's 1980 revised version especially as Morrison was a very convincing and unshakeable witness in the face of a skilled cross-questioning by the police at which I was present.

The whole story stresses the point that footprints, hair, and sightings convince no one nowadays, and only a photograph or, heaven forbid, a cadaver will be accepted as proof that the thylacine is not extinct.

The 1963–1964 expedition

By 1962 the evidence which the Fauna Board had collected over the years was sufficiently convincing for us to go to the Tasmanian government and request funds for an expedition for us to conduct a special search to try to capture a thylacine by concentrating in areas which we considered most likely to give results. The Premier, Eric Reece, hopeful of success and envisaging the publicity that this would engender, granted us £2000 to be spent on expedition costs and in employing a professional snarer to help us.

I must confess that I had, and still have, grave misgivings about what should be done whenever a thylacine is captured or photographed. A photograph is relatively easy to cope with since the area in which it was taken can be concealed, but a real live thylacine on the end of a piece of rope is an entirely different matter. We spent hours round our camp-fires talking about this dilemma but, fortunately or unfortunately, it was one we did not have to face. We knew that however remote the place might be, news of such a capture would leak out sooner or later, and once the publicity hounds got the story anything could happen.

After these discussions and subsequent representations to the Reece government, Maria Island was decided upon as a sanctuary and acquired in perhaps one of the best moves in Tasmanian conservation history, especially as it was being considered by business interests for imminent development as a tourist resort.

On this expedition to capture a thylacine we decided to use snaring as the method since we could cover more country in remote districts with greater efficiency than by using cumbersome traps which were of doubtful use for catching thylacines. There was some criticism of this method at the time on the grounds that the animal would have been killed by the snare but most of these objections were ill-founded and were made by people who knew little about snaring and still less about the methods we used. The only risk was post-capture shock which the old trappers had told us about, but we had to risk this and tried to reduce its likelihood to a minimum by going round the snares daily.

We hoped to catch the thylacine unharmed and to achieve this snares using a strong springer and a straight stick as a treadle were set very finely so that only the legs would be caught. This was the case almost invariably, except in one instance when we caught a rather astonished pademelon by the tip of the tail and he was very cross by the morning. To the best of my knowledge we did not kill any animals in the snares.

Five persons took part in the expedition which was under the field supervision of Inspector G.J. Hanlon BEM, who was assisted by senior wildlife officers Reuben Hooper and Ken Harmon as well as by a professional snarer, Ray Martin, and myself. Both Ray and his father had experience of thylacines, Ray having seen one with its cub chasing a kangaroo on the South Eldon River in 1952. His father caught thylacines for the bounty. I went along as frequently as my university duties would allow. A roster was arranged so that there were two people in the field at all times, with Ray being the regular man. The expedition was described briefly by Guiler (1966).

We started snaring at Green's Creek running a snare line down behind the dunes as far as the Pedder River as well as up into the hills behind Green's Creek. About 700 snares were set, and the daily examination of these with resetting and replacement of damaged springers took all day.

The area was difficult of access, especially in wintertime, even to a four-wheel drive vehicle. Our adventures started with the crossing of the Arthur River on the punt propelled by the muscle power of Bob Airey, and from there to Cawood the track was not too bad, although in spots it had its trials. From Temma to the south it was poor to downright bad, one boggy patch at the Dawson River being about 400 metres in every direction, especially, it seemed, down. It usually took us nearly two days to get from Hobart to Green's Creek, or to Balfour in the later stages.

The camp at Green's Creek was a tin hut constructed for the lighthouse people and how we wished that the galvanized iron sides had extended down to the floor and not terminated about 10 cm above it. When the wind blew, which was constantly, a howling draught swirled round our ankles and when it rained, which was every day, a film of water flowed over all of the floor. Living there was quite an experience and the only light relief was a trip to Sandy Cape in the Toyota four-wheel drive dodging the waves and hoping that we would not stick in the quicksands.

We caught a lot of animals but no thylacines. Between 31 October and 27 November we snared 7 brush possums, 188 pademelons, 46 wombats, 46 wallabies, 3 tiger cats, 28 Tasmanian devils, 1 brown bandicoot, 2 echidnas, 6 marsh harriers, 1 sea eagle, and 1 crow. Some of these were caught more than once. Not only does this list give a picture of the variety of animals caught but it indicates the very fine setting of the snares to catch such a small creature as a brown bandicoot.

With the proximity of Christmas and the appearance of a few intrepid holidaymakers who asked awkward questions about our activities, we moved north to Woolnorth and concentrated on the Studland Bay run where another 700 snares were set. During the time we were at Woolnorth we located a lair that allegedly had been used by thylacines in the previous winter. We set another 100 snares around this area. One of the difficulties we encountered was in getting enough material suitable for springers. We cut these at Tarraleah and carted them to Woolnorth. It is little wonder that the Van Diemen's Land Company snarers used necker snares which do not require springers. Here we again caught a variety of game: 150 pademelons and 104 devils headed the list followed by 24 wallabies, 10 tiger cats, 2 brush possum, 1 wedge-tailed eagle, 1 hawk, and 1 crow.

The party remained at Woolnorth for the summer and then took a break before heading off to Balfour, which is near Green's Creek but directly inland from Temma. Balfour used to be a mining town, first for copper and then as an open-cut mine for tin, and a service port grew up at Temma where the Whale's Head Hotel offered 'tiger' hunts as an attraction. Balfour had been abandoned for many years but sporadic small-scale mining had taken place in recent years. No one was mining while we were there and we lived in the only surviv-

The end of the Balfour search. Reuben Hooper giving advice on how to get the Rover back on its wheels.

ing shack; all that was left of the town at this time were a few bricks, the concrete slab and chimney of Rainbird's old hotel surrounded by a large patch of bracken, and six totally unexpected and unprotected deep wells. It usually rained every night, sometimes quitting by about 10.00 AM whereupon we set off on our rounds. Other days it did not stop raining and the bush was wet and miserable.

Another 700 snares were set in different types of country, mainly in the fringes of the forest around the edges of the buttongrass plains as well as on game trails through the forest. Walking trips were made around and about for reconnaissance purposes. The results were much the same as before—the same variety of game caught in some numbers but no thylacine. We did see a probable thylacine footprint close to our camp on a track past the copper mine towards the sluice face.

Balfour was a miserable place in winter. The track from Temma was rough, rutted, waterlogged all the time with water so deep in places that it dripped out of the back of the dashboard of the Land Rover as we went through, and sometimes the vehicle would stick in the mud and we had the tedious job of de-bogging it. Often we arrived back at camp late in the evening very tired and very hungry too.

The scrub was thick and even walking along the tracks which we had to make was an effort, especially in the cutting grass, which was about 2 metres or more in height and constantly wet.

The expedition persisted in the area until the end of May when general exhaustion overtook everyone and the finances as well. A vehicle accident on the road into Balfour in which three of the expedition were hurt finally put an end to what was a pretty good effort. The team worked very hard and loyally in the wretched conditions. Time has dulled these miseries somewhat, and we can joke now about the de-bogging of the vehicle and even laugh about such gruesome things as Hanlon's wallaby stew.

There is a lot of very tedious, very hard work involved in the maintenance of a long snare line. In order to set the snares we had to cut 700 springers for each locality, 700 treadles, 700 straight sticks, and make 700 buttons, plus roll the hemp for the snare and then carry the whole lot through the thick scrub into position. In addition, we then had to make replacements for all the snares the devils chewed. Each day we walked 15–20 kilometres through rough country, and I think that the sheer grind around the lines every day was the principal factor which exhausted us.

We also made field investigations aimed at determining areas for future searching. We looked at south of Macquarie Harbour, south of the Gordon River, north of the Pieman River, and south of the upper reaches of the Arthur River. None of these looked any better, or if there were any possibilities of success it was offset by the difficulties of access and of obtaining snaring material. The areas we covered included that recommended by Summers for reservation as a sanctuary where he had seen signs of thylacines. Woolnorth was a natural for a search and we could not have ignored it.

If we count one snare set for one night as giving one chance to catch a thylacine then we had somewhere in the order of 126 000 chances, and we felt we deserved better luck for all the effort we put into the project.

The West Coast (Lyell) Highway investigation

In July 1961 I went to look at a cave near the West Coast (Lyell) Highway where a nest had been discovered by Mr Hank Meerding. The cave had been occupied when first found in May 1961, as the nest was warm. The floor of the cave was of clay with sandy patches and the nest had been hollowed out of the cave floor and then lined with twigs and grass. There were claw marks on the inside made by the young as they scrambled to get in and out of the nest. The nest was quite far into the cave and was unlikely to have been made by a bird as there were no feathers lying around.

There was a lot of dung lying about on the floor of the cave and this showed that two sizes of animal had been using the cave. None of the droppings were close to the nest. There were some footprints in the sand, the clay being too hard to take an impression, but none of these gave a clear print. There were some bones in the cave, indicating that food may have been brought to the nest site. The droppings were very large to be those of a devil, some being 12 cm in length. This was an irritatingly inconclusive investigation as we do not know which species would make a nest such as this. The next day Hank and I went for a walk to the top of the Raglan Ranges and then went about 18 kilometres towards Frenchman's Cap but there were very few tracks of any animal to be seen. A visit by Hank Meerding's son in 1984 failed to show any further traces of the animal.

The second west coast expedition

An island sanctuary was first suggested by Flynn (1914) and now with the requisition of Maria Island as a sanctuary there was a suitable place where a thylacine could be released, and we decided in the winter of 1966 to make another effort to catch one. The World Wildlife Fund gave us $1000 towards the quest. Before settling on a locality we decided first to investigate for signs of a thylacine living in the area and some time was spent in this activity. Eventually we located a region near the west coast where footprints had been seen by an old man who knew thylacines and we set off to snare this region. The party consisted entirely of Fauna Board employees and we spent about three months in the area, again without success.

The country was easier to work through and we had a decent hut in which to live. It was waterproof, the walls reached to the floor, and the family of rats living behind the fireplace was either toasted when we lighted the fire or decided to seek another home.

In retrospect, I think that this expedition was run too soon after the previous major effort as the personnel still remembered too well the disappointments of that search and their enthusiasm had not had time to be renewed.

The Whyte River search

In June 1966 my attention was drawn to a possible lair in an old abandoned boiler at an old gold mine at the Whyte River. The place was discovered by Reuben Charles and he, together with Rex Davis who operated a sawmill nearby, helped me with the subsequent work.

The area is fairly thick forest which had been logged many years ago and later cleared for mining operations. It was a bit hazardous with old shafts, exploratory holes, and so on, but the access road to the mine could still be used for walking to the boiler.

A trap was constructed with a compartment in the middle which was separated from the trap cage. There were two cage compartments on either side of the middle section and these had drop doors operated by a treadle arrangement. The central section was used as a home for the live bait which was placed inside and provided with food, water, and shelter. It could not be killed by an animal in the trap compartment. The lure was a live wallaby or a hen, the latter was more productive as it provided an occasional egg.

As well as this we rigged up a camera using a somewhat updated version of our Mark II gear and replacing the G24 cameras with an 8 mm Halima camera. This Mark III version took some photographs, mostly of padememons and bandicoot, but one photographic sequence showed a blurred animal moving quickly in the lower corner of the frame. It could not be identified.

We searched the area for footprints and met with no success except for the usual devil, wombat, wallaby, and pademelon prints and those of the mill dogs. The trap caught all the usual oddments but the number of animals caught was low. No tiger cats were captured. We found on several occasions that the trap had been sprung by some animal large enought to activate the treadle near the bait while having enough of its rump under the door to stop it closing, so allowing the animal to reverse out. Some hair collected was of thylacine origin but only dog spoor was found around the cage.

The trapping continued for about a year, largely due to the enthusiastic efforts of Rex Davis, but without success.

The Griffiths–Malley expedition
This was the biggest effort made up to the time to find a thylacine and really deserved more success than it had. Jeremy Griffiths from New South Wales and James Malley of Trowutta set off in March 1968 on a search which was to some extent sparked off by Malley's discovery, described earlier in this chapter, of something which had threshed about the bush while in a snare. They walked over a great deal of Tasmania, particularly the north-east and west coast districts and even trekked to Port Davey from Macquarie Harbour.

They later established the Thylacine Expedition Research Team, being joined in this by Dr R. Brown, a Tasmanian bushwalker. In due course money was subscribed and twenty-five automatic camera units were spread widely over likely localities. In addition to all this, a Tiger Centre was established for the public to give and seek information. This brought in some useful contributions and aroused a deal of public interest.

The two leaders of the expedition did a tremendous amount of work over the four years of their quest and, if work counted, they should have got results. However, they, like all of us, got neither their photograph nor an animal, but they did collect a lot of information as they went along and enough sighting reports to encourage them in their efforts. The team disbanded in December 1972.

The principals in this expedition disagree on whether or not the thylacine is extinct, Malley being certain that it is not, and both Brown and Griffiths believing to the contrary (Griffiths *et al.* 1972). They concluded that a factor in thylacine extinction was the trappers habit of poisoning snare lines to kill devils and these baits being eaten by thylacines. However, from what we have been told of thylacines in the wild this is unlikely as they are not scavengers.

The Sayles and Tangey searches
These two, essentially solo, searches were conducted in the north west and the north east respectively and were not successful. Sayles, who was assisted by the enthusiastic James Malley, used a typically American invention called a 'varmint caller' which squawked like a wounded animal. He spent nights perched in trees puffing at this instrument and on one occasion he succeeded in attracting an animal which he could not identify but it was much bigger than a devil. He is now left with that awful hanging doubt 'What was it?' (Sayles 1980).

The use of the caller is interesting but whether it would have attracted a thylacine is another matter. Most marsupials are silent when hurt and an animal's sudden cry of pain is unknown of thylacines. It was an original approach which might well have paid off and was worth a try.

Other searches
The mystique that surrounds the existence of the thylacine has attracted people to Tasmania to search or try to search for it. Some of these individuals or groups have absolutely no experience of the Australian bush let alone the Tasmanian scene, and I fear that some of them get rather a shock when they encounter what Gould described as our 'rougher country', and the west coast impenetrable horizontal scrub. Even such stalwarts as Sir Edmund Hillary have become involved,

perhaps more to newspaper enthusiasm than by their own intentions. In Sir Edmund's case he came to Tasmania in 1960 to climb mountains in the south west with Jack Thwaites, but as soon as he expressed interest in seeing a thylacine this was interpreted by the press as the main thrust of the visit.

The American press and TV have been interested in sighting reports over the years, and in 1958 a Disney Film Unit came on a visit here and proposed to make a film on the thylacine. The team, comprising Al and Elma Milotte who had worked on *African Lion*, was diverted into more profitable channels and spent some time working on Tasmanian devils at Westbury. This was, I believe, the origin of the comic strip on the escapades of that animal. The National Geographic Society has expressed interest and some reference was made to the thylacine when a crew visited Tasmania in 1967 but they did not conduct a search or major investigation (Brander *et al.* 1968).

The World Wildlife Fund (Australia) search

The World Wildlife Fund (Australia) was launched in 1978 and as part of its initial programme decided to assist in making a search for the thylacine as one of its major efforts. A grant of $55 000 was made to the National Parks and Wildlife Service for a two-pronged approach to be made by the Service and myself.

The Service appointed Steve Smith to investigate recent sighting reports and conduct field operations at the sites, using three specially constructed automatic cameras for the purpose. He interviewed a few of the remaining people who knew of thylacines and collected some historical information and sighting reports. His results were encouraging but still no thylacine or photograph, and in the absence of any conclusive evidence he was forced to concede that the species was extinct. Details of his search can be found in his report (S.J. Smith 1980).

My own part in the search was based on a two-year field exercise using automatic cameras. I firmly believe that little is to be gained by rushing frantically about Tasmania investigating each new alleged sighting. Fleay did this in the

1940s, I did it in the late 1950s and the 1960s, and the Griffiths–Malley team did it as well, and we all got nothing out of it. Similarly, I believe that little new information will be gleaned now, most of the recent interviews are largely repetitive, although some historical snippets could still turn up.

Instead I am convinced that cameras placed in likely areas over a long period of time will bring more chance of success, and with this in mind I spent a considerable proportion of the $25 000 which was allocated to me on the construction of fifteen automatic camera units, a prototype of which had been designed for another project.

Finding a suitable area in which to place the cameras was no easy matter. The units did not require much servicing and could be left in the bush for about fourteen days, but the area had to be somewhat isolated and free from interference and vandalism. There also had to be good sighting records in recent years and if possible in a district historically noted for thylacines.

The equipment The apparatus consisted of a pulsed infra-red beam which when interrupted activated a movie camera and lights. Each unit consisted of a transmitter, sensor, control box which doubled as a carrying case, camera, and lights. The transmitter circuit was mounted on a printed circuit board housed in a plastic box (9 × 7.5 × 3.2 cm) to which a 10 cm length of 1 cm diameter steel rod was permanently attached as a mounting. Two infra-red-emitting diodes set horizontally about 1.25 cm apart protruded about 2 cm from the box. Power from a 12 volt lantern battery entered through the bottom of the box. The transmitter and sensor were mounted on steel rods using a laboratory boss head.

The sensor consisted of an infra-red-sensitive diode housed in a metal tube (15 × 1.25 cm). The face of the diode and its filter was recessed about 2 cm from the front of the tube and was enclosed between two pieces of nylex tubing and a celluloid disc. A 3 m cable, sealed at its exit from the tube with a silicone sealant, was connected to the control unit with an RCA-type plug and socket, colour coded. The control unit circuitry was constructed on two printed circuit boards mounted, together with a four-position rotary switch, on an aluminium angle chassis. The switch and a two-digit display counter were mounted on the front edge of the chassis.

The activation period for each incident was 2.5 seconds, and the number of incidents was recorded on the display counter. An inhibit period of 30 seconds was allowed between incidents to allow animals to move away. Both of these times were adjustable. Approximately eighty incidents were obtained from one film magazine and overflow of the display caused inhibition of further incident activities, resulting in saving of battery power. Further power savings were achieved by a photoelectric cell which switched off the light source during daylight.

The camera was a super-8 cine type with an f/1.2 lens and auto stop. Two brands were used, namely Kodak XL 320 and Sanko XL, five of the former and fourteen of the latter. The camera was mounted in a housing of sewer-grade PVC tubing modified to have a hinged rear door with an 'O' ring seal and safety pin together with flanges internally fitted to mount the camera, which was recessed to prevent spotting of the lens by rain. The camera, together with the lights, was mounted on a 2 m Dexion arm and connected to the control box by a 3 m colour-coded lead.

The light source initially was two 21 watt tungsten bulbs but these were replaced by one 50 watt quartz halogen lamp. Power was supplied from a 12 volt (42 amps/hour) lead acid battery. The lights were connected to the control unit by a 3 m cable.

The transmitter and sensor were placed to cover an animal trail and were no more than 4 m apart, usually much less. The alignment of the sensor and transmitter was critical and adjusted by an audio signal when not correct. The light source had to be arranged not to shine on the photoelectric cell. All the equipment was arranged so that rain would run off the boxes.

Fifteen units were constructed, each costing about $550. The transmitter, sensor, and control unit were electronically balanced and remained together but the lights and cameras were interchangeable. In practice, only about twelve units could be maintained in the field due to the problems outlined in the next subsection.

Field problems   Animal damage ranged from nil to severe. The most extreme damage was caused by cattle which kicked over the batteries, chewed most of the leads, knocked over the camera mounting, and trampled on the sensor and transmitter. However, the electronics did not suffer any damage.

Tasmanian devils chewed the leads, particularly that from the sensor. On one occasion a spider built a web and nest inside the sensor tube, and there was some trouble from ants building nests inside the camera box and body.

The units have been exposed to a considerable range of weather conditions including temperatures from −5°C to over 40°C. They have operated in areas of high rainfall (150 cm/annum) and have twice been buried under snow. Moisture in the wooden boxes caused trouble by starting corrosion at the terminals on the box, particularly around the photoelectric cell, but this was corrected by isolating the terminals from the box. Excessive humidity caused corrosion of the printed circuit. Two units were written off after animals had overturned them and water made its way into the control panel.

The units could be left undisturbed *in situ* for about ten days, after which routine maintenance was necessary and the alignment of the sensor, amount of film, and operational efficiency of the equipment had to be checked. Every twenty days it was essential to check the batteries, particularly the 12 volt accumulator.

The disadvantage of the system is the weight of the units, lights, batteries, and stand which weighed about 40 kg. A modified backpack enabled two units to be carried but when trails were suitable a wheelbarrow could be used to transport four units. The distance that the units can be carried is thus limited and in practice could not be far from a track.

Some ageing has been experienced in the electronic units, particularly the integrated circuits of the transmitter, which stop working without any warning. After three years of rather rough wear, ten units are still functional.

Field investigations   Four units were taken for field trials to Adamsfield for four weeks. Minor electrical faults were corrected and the programme started immediately (Table 8.3) when it was decided to concentrate the units in one area

TABLE 8.3   Camera unit use, 1980–83 thylacine expedition

No of. units	Location	Period	Reels of film used
7	Adamsfield	17.6.80–4.9.80	41
15	West coast	29.9.80–20.12.81	50
15	Mainwaring River	23.2.82.–20.3.82	30
10	North west	22.3.82–1.12.82	41
1	Lyell Highway	13.12.81–22.12.81	1

rather than scatter them widely. The sites were chosen in areas where thylacine sightings had been reported in recent years; I will not give full details here for security reasons as some investigations are still proceeding.

Each of the sites offered different habitat types, Adamsfield having the highest rainfall (180 cm/annum). The vegetation is wet sclerophyll forest with open sedgelands on flat marshy areas, the cameras being set on trails leading from the forest to the plains. At a later stage, the seven units were set along a track leading to Adamsfield at a higher altitude and some of the precipitation fell as snow.

The west coast area was drier, having only about 140 cm rainfall, and with more varied vegetation. Much of the area was cleared pasture land with heavily wooded gullies and surrounded by rainforest dominated by *Eucalyptus* or *Notho-fagus*. The units were located on trails leading into the paddocks.

The Mainwaring River site was located on the coastal dunes and grassy areas in the coastal scrub. There was no forest in the area, the only trees being some stunted *Eucalyptus* on the coast. Extensive sedgeland plains extend inland behind the narrow coastal belt. The rainfall is about 140 cm/annum but there is considerable salt spray from the very rough seas which roll in on the coast. The cameras were set on trails leading to the grassy feeding areas used by herbivores. The area was accessible only by helicopter.

The north west area was located inland in a region of mixed regrowth overmature myrtle forest with a 140 cm annual rainfall. The area had been logged and the camera units were dispersed along the logging tracks.

TABLE 8.4   Species photographed by the 1980–83 survey

Species	Location	Comments
Bennett's wallaby, *Wallabia rufogrisea*	All	Most abundant at west coast, uncommon at Mainwaring
Rufous wallaby, *Thylogale billardieri*	All	Very common, except at west coast
Potoroo, *Potorous tridactylus*	North west only	The only small species recorded
Wombat, *Phascolomys ursinus*	All	Common
Brush possum, *Trichosurus vulpecula*	All	Not abundant
Ringtailed possum, *Pseudochirus convolutor*	West coast	Only one record, from west coast
Tasmanian devil, *Sarcophilus harrisii*	All	Very common
Tiger cat, *Dasyurops maculatus*	North west	Filmed on only one trail; not common
Cat, *Felis cattus*	All	Most numerous at west coast

The Lyell Highway site was beside a road with myrtle forest on both sides, the rainfall was about 200 cm/annum. There were no obvious trails running through the forest, consequently only one unit was used. It was placed at the spot of a recent sighting.

Results   Although every other species of large marsupial living in Tasmania, except the forester kangaroo, was photographed, the films did not show a thylacine and the expedition must, to date, be reckoned a failure. However, a great deal of information on other species and trail utilization has been gathered and can be used elsewhere.

It was clear at all the sites that the trails were used by different species on the same night and the presence of carnivores such as the Tasmanian devil, *Sarcophilus harrisii*, and the cat, *Felis cattus*, apparently did not deter other species from using the trail. The species photographed are shown in Table 8.4.

The number of animals which were active in daylight hours was surprising, about one-eighth of all the incidents taking place during this time, some occurring at the midday period as shown by the shadows. The species most commonly found active in daylight were the wallaby, *Wallabia rufo-grisea*, the Tasmanian devil, and the feral cat. Neither walla-bies nor devils were seen in daylight in the west coast locality and it is probable that the activity is confined to the heavily vegetated areas.

On several occasions an animal using a trail could be iden-tified by individual characteristics and noticed to pass along the trail in the late evening or early part of the night and return later or next morning, presumably going from its rest-ing to its feeding site.

Discussion   The frequency of occurrence of a species on the films does not give an index of its abundance in an area. For example, the thylogale, *Thylogale billardieri*, is abundant in the west coast area but did not appear frequently on the films (Table 8.4). Its absence is related to the siting of the units in places not used by the species.

The films showed a quite unsuspected abundance of brush possum, *Trichosurus vulpecula*, at the west coast site where trapping has taken place for many years, and it was surpris-ing to find them at the Mainwaring River as the scrub is very limited and bounded by extensive heathlands unsuited to brush possums. Hocking & Guiler (1983) found this species was uncommon in inland west coast rainforest and heathland country. The ringtail possum, *Pseudochirus convolutor*, had not been recorded at the west coast area until this survey.

The feral cat was common in all localities, except the Mainwaring River where only two were recorded. Feral cats are now widespread throughout Tasmania and must be clas-sified as an established part of our fauna.

The most recent search has been in an area where a National Parks ranger sighted a thylacine early in 1982. He saw the animal in a spotlight for about 2 minutes and this must be regarded as a positive sighting. I transferred my cameras to the area, where a National Parks ranger operated them for twelve months, alas with negative results, and we managed to keep the matter quiet until December 1983. This search was carried out in full co-operation with the National Parks Service and with government support, and I do not wish to say anything further about it at this stage.

Mr P. Wright of Mole Creek announced in January 1984 that he intended to organize a three person expedition to the central plateau–Western Tiers area near Mole Creek, establishing a base camp for a bushman and a zoologist. The expedition will use automatic cameras in an attempt to get a photograph. According to Wright he already has fifty reports of alleged sightings in the past two years, only eight or nine of which he would take seriously. He also claims to have collected bones and a plaster footprint. His intentions are to make a documentary film of the expedition (*Saturday Evening Mercury* 1984). The hills above Mole Creek would appear to offer as good a chance of a photograph as anywhere, as the region offers good thylacine habitat and from the historical records it has always done so.

Further areas are being investigated as promising sites for the location of the cameras and I hope to continue in 1984. Although the World Wildlife Fund (Australia) grant has been expended, further support has been obtained and I can carry on for another short time.

9

Mainland 'tigers'

From time to time reports have appeared in newspapers and magazines of unidentified animals seen at various places on the mainland of Australia. Some of them are highly fanciful, and a photograph alleged to be of a thylacine which appeared some years ago was of very doubtful authenticity, but others are interesting and worthy of consideration, where clearly the observer had seen something which could not be related to animals known in the neighbourhood.

Perhaps the most famous animal of all is the bunyip, only rivalled by the legendary mountain lion which was reputed to have been liberated by a US serviceman in the Grampians during the 1939–45 war. The story goes that a US airforceman brought six cougar cubs with him as mascots and when the Australian authorities found out about this they ordered their destruction but the serviceman liberated them. Another rumoured wild feline was the Emmaville (NSW) black panther which was shot and turned out to be a dog. Coleman (1976) has recorded a fine collection of these rather elusive creatures.

It is not my intention to review all the tales which have

appeared in magazines, newspapers, and journals, but I have selected a few incidents, and record some others which have not been published before collected from people who have written to me.

In this chapter we should bear in mind that thylacines have not been recorded from continental Australia since about 3000 years ago. It is unlikely that any have survived, particularly as there are no reports from tribal Aborigines of their existence in recent times.

South Australia

There have been reports of sightings of strange animals over many years now from South Australia as well as from the Nullarbor Plain and nearby Western Australia. Some of them give accurate descriptions of a thylacine-like animal such as this report from the *Adelaide Advertiser* of 2 November 1962:

> A strange animal sighted on three occasions about a year ago has been seen on the outskirts of Port Macdonall. Last night Ian McRae, about 19, of Port Macdonall, saw the animal on the western outskirts of the town. He described it as being about the same size as an Alsatian dog, light brown in color and with black stripes running across its back.

McRae had obviously seen a most unusual animal and also there is little doubt that he would not be predisposed towards imagining it to be a thylacine.

This report was followed up by the South Australian Field Naturalist Society and Mr A. Cockington of the Society interviewed men who claimed to have seen the animal. One rabbit trapper said that he saw the animal near a waterhole on several occasions and once saw two together. He is reputed to have wounded one with a .22 rifle bullet and followed it for 5 miles (about 8 km) before losing it. He was offered a substantial reward for a carcase by a French professor, the correspondence having been seen by Mr Cockington.

Mr Cockington sent me plaster casts of footprints and photographs of spoor. The claws of the animal were very big, so much so that I doubt if it was the footprint of any animal we know of today, particularly as the claws pointed straight down from the toes in a very unnatural position which would have been uncomfortable for the animal when walking on hard ground. One of the photographs shows footprints which are suggestive of those of a large domestic cat. In general I found this evidence very inconclusive, the casts of the spoor as is so often the case were not positively identifiable. It is generally known that feral cats sometimes grow to a very large size, and perhaps with poor light and fleeting visibility the viewers believed they had seen a most unusual animal.

Another report from South Australia also tells of a thylacine-like animal: 'Near Moonta on 14 March 1981 at 8.30 in the morning I saw a ginger-coloured animal with deep black stripes on its back.' (Mrs D.J. Kaye pers. comm.). The animal was described as mangy-looking but the viewer was certain that it was not a dingo.

Some of the reports emanate from expert observers such as that by Lighton (1968) who was a government stock inspector; he saw an animal believed to be a thylacine near Naracoorte, and observed it for a quarter of an hour (*Launceston Examiner* 1968).

Mr Huon Johnston who lives in Sydney did not report the following incident and thought little about it until he saw a TV 'World Around Us' programme which contained a segment on the thylacine, when he telephoned me to tell me of it. In May 1976 Mr Johnston was travelling by horse across the Nullarbor Plain. He and his wife formed two of a party of five, one being a child. They were at Nullarbor Station and went out early to look after the horses when they saw a thylacine. Mr Johnston said he clearly saw the stripes and the party was close enough to distinguish its bull-terrier-like head.

Victoria

Some reports have come to my notice from Victoria, and here again a few have the ring of authenticity about them.

Three persons saw a thylacine-like animal crossing the road at night on two separate occasions, in January 1970 and in February 1973. The location was near Kallista in the Dandenong Ranges. The observers were not emphatic that the animal was a thylacine but were certain from the tail, its size, and its ears that it was not a fox. The persons concerned had lived in Africa and were aware of the inaccuracy of observations made at night (McAdam 1973).

This report was preceded by that of McCance (1970) who recorded two sightings of an unknown animal of thylacine-like form near Kalorama. A further sighting from the Dandenongs was recorded by Spicer (1978) who saw what the group was convinced was a thylacine in the afternoon while driving from Yea to Healesville. She reported the matter to a Fisheries and Wildlife officer who only commented that if this had been seen in Tasmania he would have been out looking for it. And in 1978 Mr W.S. Mould (1984) saw what he believed to be a thylacine on Range Rd, Olinda. He was certain it was not a fox since he had grown up in fox-hunting country in England.

Another Victorian sighting of a thylacine-like animal was at Lang Lang, 100 km south-east of Melbourne. Three men in a car saw the animal on 26 November and again on 27 November 1979 about 3 km from the town (*Hobart Mercury* 1979).

Yet another sighting was from the Cape Liptrap-Point Smythe area, where a family obtained a clear view of an animal at about 250 m distance and they are convinced that they saw a thylacine (Eastwood 1981).

An earlier report (Lyon 1972) told of an animal seen in 1967 at Lake Victoria and the viewer was able to give a very detailed description of it. She was informed that several such sightings had been made that year. Yet another Gippsland sighting was made near Lake Buchan (Mackieson 1981).

Other sightings have been reported from Tallangatta in 1971 (*Weekly Times* 1971), and from Meningie, Bordertown, and Mt Gambier (Packer 1970), the last town being in South Australia. The Western District sightings have been made at intervals since 1964.

Western Australia

The region of south-western Western Australia and the Nullarbor near Nannup have been the scenes of sightings of unknown animals for a number of years, but in spite of the efforts of such enthusiasts as Ian Officer of Benger no animal has been captured or positively identified. Most of the sightings have been in the south west, no less than twenty-two occurring in one year (Carmody pers. comm.).

This event took place at Exmouth in 1980:

A very clear view of a dog-like animal, grey-brown in colour with very marked either dark-brown or black transverse stripes running across the rump and a thin tail. It was about 20 inches high. The viewers were able to watch it move slowly into scrub about 25 yards away. (Buckingham 1980)

Queensland

There have been reports for years emanating from northern Queensland of an animal, be it a marsupial cat, striped cat, leopard, or whatever, living in the more remote parts of the bush. Troughton (1941) considered the rumours to be of sufficient importance to include them in his standard work on the mammals of Australia.

One of these alleged animals was killed at Craignish and photographs of it were published (*Maryborough Chronicle* 2 February 1983) which showed an animal with large canines, poor but rather long fur, and a very long tail which was approximately equal to the length of the head and body. The creature was identified by its dentition as a dog by Miss J. Plunkett and I believe that she was quite correct. The tail is long and not dog-like but there is even less resemblance to a thylacine, a cat, or other legendary creature.

Northern Territory

I have only one report of a sighting of a thylacine-like animal from the Northern Territory:

A bull catcher, G. Parry, saw a pack of scavengers among which was an animal which looked like a dingo but had a number of rings around its body. The event occurred in September, 1983 and was investigated by the Northern Territory Conservation Commission. Their initial reaction was that it would be unlikely to be a thylacine. (*Hobart Mercury* 1983)

New South Wales

There have been reports of a strange animal in the Glen Innes region of northern New South Wales but little has been brought to light about this animal. It was first sighted near Inverell and has been seen at various other places, even as far south as Armidale. A photograph of it was published in the *Sunday Telegraph*, 8 August 1976, and clearly no matter what else it may be it was not a thylacine, as the tail was carried arched forwards over the back. These animals were alleged to hunt in packs and one pack of eight was reported to have killed a number of sheep in the vicinity.

R. Gilroy of Katoomba has been collecting thylacine reports from New South Wales for many years and claims to have seen a thylacine at Blackheath in the Blue Mountains in December 1972. He also has plaster casts of footprints from Wallangambi which he believes are those of a thylacine (*Hobart Mercury* 1984)

Comments

I do not vouch for the authenticity or otherwise of these reports, which are only a few among the mass of sightings of strange animals, some resembling thylacines. However, sufficient information has been presented to show that there may be an animal, particularly in Western Australia, South Australia, and Gippsland, which does not fit the general description of animals known in the vicinity.

It is difficult for inexperienced observers to identify an animal that is seen for only a few fleeting seconds in car headlights at night, but some reports are from persons who saw, in daylight and for long enough to have a good look, a creature which they firmly believe to have been a thylacine.

Although thylacines were not seen very frequently in Tasmania even during the time of their relative abundance, I think that if they were still living today on mainland Australia we would have more definite descriptions of sightings and that the tribal Aborigines would surely know of their existence. It would be a great shock to all of us in Tasmania if a thylacine turned up in the mainland, but I consider the chances of finding one are better here in Tasmania where at least we know they have existed into recent times.

10

Retrospect and prospect

We have seen that the information available to us on the ecology, behaviour, and general biology of the thylacine is so meagre and sometimes contradictory that we have to resort to some degree of conjecture as we try to piece together the life of this intriguing animal.

It was clear by 1910 that the thylacine was already scarce and was what nowadays we would call a threatened species. The point has already been made that not only Australian museums, zoos, and universities of the period but also overseas institutions should have made more effort to collect whatever material was available and make the fullest use of it. Their failure to do so can only remain as a black mark against them as, with their experience of other species which had slipped into oblivion, surely it was apparent to them that here was another mammal well down the slide to possible extinction. It was not until concern for the species was expressed by the Fauna Board in the early 1930s that any steps were taken to investigate the problem. Overseas institutions have, in general, shown little interest in the future of the thylacine and have provided only minimal support for field studies.

The end result of this neglect is that today the biologist has an uphill task with no nucleus of thylacines, not even one positive recent sighting at the time I write, and so little information to help formulate some sort of management programme for the animal's recovery.

The assessment of the value and authenticity of the 'tiger tales' is becoming more difficult. I was fortunate in the late 1950s and early 1960s when I could refer to and talk with men who had actually caught thylacines, and hear their opinions of the tales related to me. Nowadays there are no such referees and some tales are, to say the least of it, far fetched and highly coloured.

There is an increasing tendency to believe that thylacines were ferocious beasts and some of the stories (not in this book) emphasize this point. The historical accounts all describe the nature of the thylacine as docile, giving up when snared, not frightened of people and certainly seldom showing any aggression towards them. Many recollections have been embellished with the years. I believe that there still exists a wealth of 'tiger lore' which could be recorded, but how much easier that would have been in the 1920s than now.

With the advent of white settlement, farms were developed and areas cleared thus providing additional grazing for herbivorous marsupials in the pastures and these species increased rapidly with the new food resources available to them. An example of this can be seen today at Mt William National Park where very high macropod populations are to be seen on the former grazing land in the Park.

With this change thylacines had an abundant food supply, and in addition a predator and/or competitor had been removed with the cessation of roving by Aboriginal tribes.

Suddenly, not only did thylacines have abundant macropods for food but a new animal was introduced to their diet, namely sheep. The sheep were easily caught and even the ageing members of the thylacine population now had a source of food which might extend their life span for another year or two of breeding.

Numbers

There is general agreement that thylacines were uncommon in the early colonial era, commencing with Harris' comment that they were rarely seen, and continuing through various authors and in various forms as far as Gould (1863) who wrote that they were facing extermination. Hull (1871) and Silver (1874) wrote of thylacines as formerly committing havoc amongst sheep flocks, implying that they no longer did so, probably because they were scarce. The Van Diemen's Land Company records that are available show that few thylacines were killed prior to 1874 but from that date until 1904 approximately five were killed each year.

In other parts of Tasmania the killing of sheep by thylacines probably had increased considerably by about 1884 as the protests by farmers about this activity became louder at this time resulting in petitions to parliament. However, assuming that the farmers' complaints were well founded even if their claims of sheep losses were exaggerated, we can conclude that somewhere between 1875 and 1886 a change had taken place in the relationship between sheep and thylacines resulting in increased predation and we can only speculate now on the probable cause or causes of this new situation. If the sheep numbers in Tasmania had increased greatly in the 1880s it would be expected that sheep would be more liable to predation, but this was not so as the distribution of sheep in 1886 was little different from that of ten years earlier, and sheep were not pastured in large numbers throughout the year in those areas which subsequently yielded large numbers of thylacines. In fact, there was an agricultural depression about 1884–90 and land alienation had declined with few new pastures being developed in remote areas.

That the increased sheep mortality was the result of increased thylacine predation, which could have been due to a substantial part of the thylacine population in Tasmania simultaneously turning to sheep killing, is unlikely and can be discarded as an explanation. It could also have resulted from a movement of the sheep-killing proportion of the thylacine population into sheep pastures and the concentration of their efforts on bigger and more accessible flocks. The trapping data show that this did not take place as thylacines were caught in the places where they had always been found.

I think that it is much more likely that there was an increase in the thylacine population at this time, perhaps in particular amongst those thylacines which were living in or near settlements, and if there was an increase then it would be expected that this would lead to an increased sheep mortality. The theory that there was an increase in the thylacine population at this time is supported by the figure of some 6000 thylacines (which includes those recorded by Laird 1968) killed in the bounty period. This harvest could not have been sustained by the population levels described by Gould and other authors and is quite irreconcilable with the descriptions of the earliest colonists who stated that thylacines were rare and little seen.

The bounty records show that thylacines were caught around areas which had been settled for a long time, for example Stevenson's captures at Blessington and the east coast ones, and even at Broadmarsh and Mountain River which are close to major population centres. In addition, numbers of them were caught in less developed regions such as the central highlands.

The Van Diemen's Land Company records show that sheep losses due to wild dogs were higher than those attributed to thylacines. In the early days the Company had a lot of trouble with packs of wild dogs and we can presume that dogs were killing sheep on other properties. In 1836 dogs were reported as being a menace, hunting in packs even in towns, and wild dogs prowled the state and menaced the prosperity of farms. Dogs were such a nuisance even in Hobart that they were rounded up and taken out and dumped in the middle of the river, and for some time there were complaints of dead dogs on the beaches (Goodrick 1977).

The new settlers were having problems protecting their flocks against both thylacines and dogs and some relief was sought from at least one of the predators in petitioning for the bounty scheme. It would not have been practical to have introduced a bounty scheme for dogs as the canine population of the state would have been endangered, including useful working animals such as cattle and sheep dogs as well as household pets.

It is part of Tasmania's tragedy that the government, in bringing in the bounty payment to get rid of thylacines from the areas where they were causing high sheep losses at this particular time, offered a price which was high enough to be an incentive for many trappers to go out into the rugged areas and seek out the species. It should be remembered that at this time there was a general depression in the economy of the colony. If the snaring had been restricted to the periphery of the farms then only a proportion of the thylacines would have been eliminated. However such management concepts were not considered in those days.

I believe that the rapid increase in thylacine numbers which took place in the 1880s probably continued for at least the first ten years of the bounty period when catches of 200–300 per annum were taken. Rapid increases in numbers are characteristic of cyclical population changes, as are rapid declines which also were a feature in the thylacine population of Tasmania. The rapid increase in this case was facilitated by the abundance of food, while the bounty system played no small part in the decline. Thylacines around farms had had abundant food since early in white settlement, and the fact that it took over half a century for the increase in population to take place is the reason I believe that the population was responding to a cyclical change rather than directly to adequate food supplies.

Thylacine numbers remained sufficiently high to yield about 100 animals per year in bounty payments until 1908 when the species suddenly became rare. Significantly, this occurred over all of Tasmania at the same time.

If thylacines had been hunted to extinction it would be expected that they would have vanished first from those places where they had been persecuted for the longest time but, in fact, there is no evidence to suggest that this was so. Woolnorth, where they had been hunted since 1835, still produced thylacines as did the east coast. It is true that the species had disappeared for many years from the closely settled areas such as Coal River but this was due more to habitat change and disturbance than to hunting.

The sudden collapse in thylacine numbers about 1908 from disease is typical of a cyclical population crash and is supported by the evidence of Burbury and others that they died of a distemper-like disease. We have seen that pleuropneumonia was the cause of death in a population crash in the ringtail possum around 1952.

In putting forward this new hypothesis that the thylacine population increased substantially about 1875 due to a cyclical change I am not able to suggest any periodicity to the cycle. We have no evidence that such events occurred before white settlement nor have we any suggestion that the population has shown any signs of recovery, although S.J. Smith found that there has been an increase in the number of alleged sightings since 1970. However, this increase may be due to greater publicity, more public awareness, and better reporting and recording facilities than to an actual increase in thylacine populations. Nevertheless, it is encouraging.

We must remember that the other Dasyures suffered a similar decline, probably from the same disease, and that the Tasmanian devil and the native cat have largely recovered in numbers and are now abundant, especially the Tasmanian devil. The tiger cat is slowly recovering but is not yet abundant. The best documented of these cycles is that of the Tasmanian devil, which remained scarce until about 1945 when it reappeared in some of its old haunts and by 1960 it was so numerous as to be a pest in some parts of the country (Guiler 1982).

There are fewer thylacines now than at any time since white settlement but if a population cycle exists then we could expect the numbers to increase at some future time and to do so rapidly, assuming that the numbers have not dropped below a recovery level. It is to be hoped that by about the end of this century there will be thylacines to be seen in at least some parts of Tasmania.

Investigation

The techniques for investigating if thylacines are in an area have become refined in recent years but still leave a great deal to be desired. The basic equipment required is exactly the same as it was in the time of Fleay and Fleming—a strong back and a good pair of legs. Investigators must be prepared to walk miles searching and carrying whatever gear may be necessary to achieve their aim. Buckets of sand have to be carried for up to several kilometres before it is spread at a constriction in a trail or at a fallen log or other suitable spot where a footprint might be obtained.

The development of reliable automatic cameras has made it possible to keep a trail under surveillance for months if necessary, but before putting the cameras in position it is essential to be reasonably sure that a thylacine could be living in the locality. This can be determined by recent reported sightings, identification of scats, and the recognition of spoor and kills.

Although we can assume that the thylacine has a range, we do not know the range's area or shape, but from trappers' reports we know that it is not extensive. Thus if a print is found at site A, circles of various radii can be drawn using the print as their centre. This can then be repeated at the sites of other evidence, and where the circles overlap should be theoretically the best places to locate the cameras.

That is the theory of thylacine hunting but the practice doesn't quite work out that way. The animals are often sighted from a road, and the surrounding country may be thick forest or open plains in which the chances of obtaining tracks or spoor are poor. For example, the country around Rossarden where we searched was open forest with hard ground leaving no trails and was so open that animals could move freely in any direction.

Rehabilitation

We have so little solid biological knowledge of the thylacine that it is almost impossible to devise a management plan to assist in the rehabilitation of the species. The best that we can do at the present is to leave them to their own devices while at the same time ensuring that they are totally protected and left undisturbed. In order to learn more about thylacines and their habits it would be essential to have a population that can be studied and observed, and just now this does not look like being the situation for a long time yet. Only after study shall we have enough information to enable us to help the recovery of the species.

The feeding habit of the thylacine demands live prey and the only practical way at the present time that we can ensure that a population of thylacines has adequate food is by encouraging the macropod population to build up by providing good pasture.

So far I have not mentioned the greatest danger of all to any wild mammal living in Tasmania. This is bushfire. Wild bushfires are a common occurrence in Tasmania and these either kill or drive away all the animals in the affected area. It is not known whether thylacines can quickly re-establish their ranges in a new district, but if other thylacines are already there then territorial problems may arise. I have had two areas in which I have been working devastated by fires and there is always the danger that any rehabilitation or study area could easily be wiped out in this way.

I do not think that it is appropriate at this stage to try to catch a thylacine. Many expeditions and searches have been carried out using snares and cage traps but without success and with much wasted effort. Besides, I now consider that the risk of the thylacine being killed in the snare or dying from post-capture trauma is too great, and this would cause a furore from all sections of the community and bring those connected with the event under very severe criticism.

The public, often stimulated by the media, seems to be more than anxious to have positive proof of the existence of the thylacine and the only acceptable evidence will be an authentic photograph. This in itself would not settle the matter—there would be those who would go looking for the animal so that security for it and any others around would be difficult. Its territory might be disturbed and this could have a detrimental effect on the animal.

If we are fortunate enough to locate a thylacine or thylacines the ideal is to leave them where they are and to provide them with security and protection from interference. This immediately raises administrative headaches. If the animals live on crown land the problem is a matter of co-operative action between the government departments concerned. However, if the animals are found on private land the difficulties, both legal and administrative, become greatly multiplied and a great deal of negotiation will be necessary to assure the safety and security of the thylacines and to enable investigations to be carried out without interference from the public, however well meant.

I would hope that the thylacine would not be found in an area in which forestry, mining, or hydroelectric activities were being carried out, or in any area having any potential for these purposes. These interests have been treated preferentially by a series of Tasmanian governments, and consequently a watch must be kept to ensure that these or any other developments do not interfere with a possible thylacine recovery.

I have noted in Chapter 2 that not many of our national parks supply all of the requirements for a thylacine reserve, but we are fortunate still to have in Tasmania places in parks, or on crown or privately owned land, where thylacines could still live and breed and re-establish themselves. The areas may not be ideally suited to thylacines, but even if we had a region with ideal conditions—large, well stocked with game, isolated, under state control, and with no human activity— there is no guarantee that thylacines would prosper if introduced there.

We still do not know the size of the home range of a thylacine and the factors that control its shape. During the course of our investigations in every instance, except at Woolnorth, very little supporting evidence of the existence of thylacines was obtained after the initial report. It would seem as if the animal had moved out of the area as a result of the disturbance caused by our activities, and indeed this seems to have been the experience of most of the other searchers. We know that thylacines will stay in an area for some considerable time but we do not know whether they remain there as a single animal or family group, or whether they stay there for mating, breeding, rearing young, or hunting. The Woolnorth diaries show that thylacines could not easily be driven from their haunts, and although there are stories from the early days that thylacines moved around, trappers have said that once an animal was located in an area sooner or later with persistence it would be caught.

In December 1983, to commemorate *Condor's* win in the Sydney-to-Hobart Yacht Race, Mr Ted Turner of the United States announced a reward of $100 000 for any positive and confirmed evidence of the existence of the thylacine. Such rewards, however well meant, encourage searches and expeditions of various degrees of competence and must lead to disturbance of the remaining thylacines. At best, an expedition will turn up some evidence whereas at worst the result will be a dead thylacine which is the last thing that anybody, particularly the sponsors, would wish to see.

In any case these private expeditions are an administrative nightmare as there is no assurance that security for the thylacine and its habitat will be obtained before the news is released to the media. The spate of publicity following the Turner offer led the National Parks and Wildlife Service to issue a warning that catching, trapping, or shooting a thylacine carried a maximum penalty of $5000 and/or six months' jail.

It never ceases to surprise me that since 1936 it has been lamely accepted that the thylacine was extinct or nearly so, even in the face of persistent sighting reports, some of which will stand considerable critical examination. This is a Tasmanian tragedy and it is disappointing that no world fauna body has sponsored a thorough search for this, the rarest of the world's mammals.

In this book you will have gathered that at all times and in spite of many dissenters I assume that the thylacine still exists. I believe that it may still increase in numbers, but that its best chance of recovery is to be left to itself in the bush, disturbing it and its environment as little as possible.

Now that we have reached the end of the tale we are left with almost as many conflicting views as we began with.

APPENDIX

Thylacine bounties claimed in various parts of Tasmania during the period 1888–1909

North-east Tasmania

Back Creek	1	Gladstone	30	Pioneer	5
Boobyalla	2	Golconda	20	Piper's River	15
Bridport	1	Jetsonville	3	Winnaleah	1
Cape Portland	6	Karoola	3	'North east'	2
Dilston	7	Lefroy	4	**Total**	**100**

Eastern massif

Avoca	25	Legerwood	6	Ringarooma	29
Ben Lomond	6	Legunia	1	Royal George	22
Blessington	47	Leipzig	43	St Helens	12
Breadalbane	3	Lilydale	5	St Leonards	1
Deddington	18	Mangana	9	St Pats River	7
Evandale	7	Mathinna	12	St Pauls River	1
Fingal	40	Moorina	1	Scottsdale	5
Goshen	3	Myrtle Bank	1	Springfield	6
Hall's Track	2	Nile	3	Turner's Marsh	4
Lebrina	8	Patersonia	1	White Hills	29
				Total	**358**

East coast

Antill Ponds	31	Jerusalem	4	Runnymede	1
Bagdad	1	Kellevie	2	St Mary's	19
Bicheno–Cranbrook	91	Koonya	7	Sandford	1
		Lisdillon	22	Snake Plains	2
Black Brush	1	Little Swanport	23	Stonehenge	30
Bream Creek	5	Mt Seymour	17	Swansea	17
Brighton	1	Nubeena	4	Triabunna	7
Broadmarsh	11	Nugent	28	Tunbridge	12
Campbell Town	36	Oatlands	3	Tunnack	19
Dunalley	6	Orford	2	Woodsdale	6
Eastern Marshes	5	Rhyndaston	5	York Plains	3
Epping	1	Riversdale	7	Parattah	20
Forcett	1	Ross	60	'East coast'	7
Green Ponds	3			**Total**	**521**

South

Castle Forbes	1	Hythe	1	Mountain River	2
Catamaran	1	Kettering	2	Northwest River	5
Garden Is. Creek	3	Lady Bay	1	Port Cygnet	7
Geeveston	10	Leprena	1	Raminea	6
Glen Huon	3	Long Bay	1	Recherche Bay	4

| Hastings | 2 | Middleton | 1 | Surges Bay | 2 |
| | | | | Total | 53 |

Central highlands

Barrington	6	Fentonbury	10	Mike's Hill	1
Beulah	1	Glenora	34	Mole Creek	14
Bishopsbourne	1	Gowan Brae	2	Natone	2
Blackwood	23	Great Lake	2	Needles	1
Creek		Gretna	16	New Norfolk	11
Bothwell	18	Hadspen	3	North Motton	3
Bracknell	3	Hagley	1	Ouse	37
Carrick	2	Hamilton	24	Plenty	3
Chudleigh	7	Hamilton-on-	8	Railton	9
Claude Road	1	Forth		Sheffield	6
Cluan	1	Hollow Tree	1	Tyenna	4
Cressy	17	Interlaken	12	Uxbridge	4
Derwent Bridge	235	Lachlan	1	Victoria Valley	11
Deloraine	9	Liena	6	Westbury	26
Dunorlan	1	Longford	40	Wilmot	17
East Meander	30	Lowe's Bridge	4	'Highlands'	2
Ellendale	11	Macquarie Plains	2	Total	688
Exton	1	Middlesex Plains	4		

West

| Royenrine | 1 | Strahan | 4 | Waratah | 12 |
| | | | | Total | 17 |

West Tamar

Beaconsfield	2	Glengarry	9	Rosevale	5
Black Sugar Loaf	6	Holwell	1	Rosevears	2
Bridgenorth	1	Parkham	16	Winkleigh_	1
Elizabeth Town	25			Total	68

North and north-west coast

Black River	11	Montagu River	32	Stanley	83
Duck River	12	Penguin	3	Stowport	1
Flowerdale	2	Ridgely	1	Sulphur Creek	1
Forest	7	Sister's Creek	18	Ulverstone	9
Latrobe	1	Smithton	18	Wynyard	4
Melrose	1	Somerset	9	Woolnorth	84
				Total	236
		Grand Total	**2110**		

References

ALLEN, H.   1958   Letter to the author.
ALLPORT, M.   1868a   Remarks on Mr. Krefft's Notes on the Fauna of Tasmania. *Pap. Proc. Roy. Soc. Tasm.* 1868, 33–6.
ALLPORT, M.   1868b   Notes on the fauna of Tasmania. Appendix to *Pap. Proc. Roy. Soc. Tasm.* 1868.
ANDERSON, H.H.   1905   *A Geography of Tasmania.* Sydney: William Brooks.
ANGAS, G.F.   1862   *Narrative of Australia:* a popular account. London: Society for Promotion of Christian Knowledge.
ANON.   1980   Suche nach dem Beutelwolf mit automatischen Kameras. *Frankfurter Allgemeine Zeitung,* 25 June.
ARCHER, M.   1971   A re-evaluation of the Fromm's Landing thylacine tooth. *Proc. Roy. Soc. Vict.* vol. 84, 229–34.
ARCHER, M.   1974   New information about the Quaternary distribution of the thylacine (Marsupialia, Thylacinidae) in Australia. *J. Roy. Soc. Western Aust.* vol. 57, 43–50.

ARCHER, M. 1976a The Dasyurid dentition and its relationships to that of Didelphids, Thylacinids and Borhyaenids (Marsupicarnivora) and the Peramelids (Peramelina; Marsupialia). *Aust. J. Zool. Supp. Ser.* vol. 39, 1–34.

ARCHER, M. 1976b The basicranial region of marsupicarnivores (Marsupialia), interrelationships of carnivorous marsupials, and the affinities of the insectivorous marsupial peramelids. *Zool. J. Linn. Soc. Lond.* vol. 59 (3), 217–322.

ARCHER, M. 1978 The status of Australian dasyurids, thylacinids and myrmecobiids. In M.J. Tyler (ed.) *The Status of Endangered Australian Wildlife*, pp. 29–43. Roy. Zool. Soc. S. Aust.

ARCHER, M. 1982 A review of Miocene thylacinids (Thylacinidae: Marsupialia), the phylogenetic position of the Thylacinidae and the problem of apriorism in character analyses. In M. Archer (ed.) *Carnivorous Marsupials*, vol. 2, pp. 445–76. Sydney: Roy. Zool. Soc. NSW.

ASDELL, S.A. 1946 *Patterns of Mammalian Reproduction.* New York: Comstock.

BACKHOUSE, J. 1843 *Account of a visit to the Australian Colonies.* London: Hamilton, Adams.

BARNETT, C.H. 1970 Talocalcaneal movements in mammals. *J. Zool. Lond.* vol. 160, 1–7.

BEDDARD, F.E. 1891 On the pouch and brain of the male thylacine. *Proc. zool. Soc. Lond.* 138–45.

BEDDARD, F.E. 1903 Exhibition of and remarks upon sections of the ovary of the thylacine. *Proc. zool. Soc. Lond.* 116.

BELL, E.A. 1965 *An Historic Centenary.* Hobart: Mercury Press.

BELL, E.A. 1967 Article in *Australian Womens' Weekly*, 10 May, p. 10.

BENSLEY, B.A. 1903 On the evolution of the Australian Marsupialia; with remarks on the relationships of the marsupials in general. *Trans. Linn. Soc. Lond. (Zool.)* vol. 9, 83–217.

BINKS, C.J. 1980 *Explorers of Western Tasmania.* Launceston: Mary Fisher Bookshop.

BLAINEY, G. 1966 *The Tyranny of Distance.* Melbourne: Sun Books.

BOARDMAN, W.   1945   Some points in the external morphology
                              of the pouch young of the marsupial
                              Thylacinus cynocephalus. *Proc. Linn. Soc.
                              NSW* vol. 70, 1–8.
BOWLER, J.H.,   1976   Late Quaternary climates of Australia
  HOPE, G.S.,   and New Guinea. *Quaternary Res.* vol. 6,
  JENNINGS, J.N.,   359–94.
  SINGH, G. &
  WALKER, D.
BRANDER, B.,   1968   'Australia'. Washington: Nat. Geog.
  HARRELL, M.A.   Soc.
  & HOLTHOUSE, H.
BRANDL, E.   1973   *Australian Aboriginal Paintings in Western
                              and Central Arnhem Land.* Canberra:
                              Aust. Inst. Aboriginal Studies.
BRETON, W.H.   1834   *Excursions in New South Wales, Western
                              Australia and Van Diemen's Land.* Lon-
                              don: R. Bentley.
BRETON, W.H.   1846   Excursion to the Western Range. *Tasm.
                              J.* vol. 2, 121–41.
BRETON, W.H.   1847   Description of a large specimen of *Thy-
                              lacinus harrisii*. *Tasm. J.* vol. 3, 125–6.
BRIGGS, A.L.   1961   Letter to the author.
BROWN, R.   1973   Has the thylacine really vanished? *Ani-
                              mals* vol. 15 (9), 416–9.
BROWN, R.   1983   Is there hope for our tiger? *The Tasma-
                              nian Mail* 16 August, p. 8.
BUCHMANN,   ⁻   1977   Behaviour and ecoogy of the Tas-
  O.L.K. &   ⁻   manian devil *Sarcophilus harrisii*. In
  GUILER, E.R. ⁻   Stonehouse & Gilmore (eds) *Biology
                              of Marsupials*, pp. 155–68. London:
                              Macmillan.
BUCKINGHAM, G.   1980   Letter to the author.
BUNCE, D.   1857   *Australasiatic Reminiscences.* Geelong:
                              J. Brown.
BURBURY, F.   1953   Letter to the author.
CALABY, J.H.   1971   Man, fauna and climate in Aboriginal
                              Australia. In J.D. Mulvaney & J. Golson
                              (eds) *Aboriginal Man and Environment in
                              Australia.* Canberra: ANU Press.
CAVE, A.J.E.   1968   Mammalian olecranon epiphyses. *J.
                              zool. Lond.* vol. 156, 333–50.
COLEMAN, R.   1976   There's a strange, strange beastie out
                              there. *Melbourne Herald* 20 March, p. 29.
COLLINS, L.R.   1973   *Monotremes and Marsupials.* Washington:
                              Smithsonian Institute.
COOK, D.L.   1963   *Thylacinus* and *Sarcophilus* from the Nul-
                              larbor Plain. *Western Aust. Nat.* vol. 9,
                              47–8.

CORRESPONDENT. 1924 Article on the killing of a tiger at Waratah by C. Penny. *Weekly Courier* 17 January, pp. 26, 46.

CRISP, E. 1855 On some points relating to the anatomy of the Tasmanian Wolf (*Thylacinus*) and of the Cape Hunting Dog (*Lycaon pictus*). *Proc. zool. Soc. Lond.* 1855, 188–91.

CROWTHER, W.L. 1883 Correspondence. *Proc. zool. Soc. Lond.* 1883, 252.

CUNNINGHAM, 1882 Report on some points in the anatomy D.J. of the thylacine (*Thylacinus cynocephalus*), cuscus (*Phalangista maculata*) and phascogale (*Phascogale calura*) collected during the voyage of H.M.S. Challenger in the years 1873–6. *Challenger Rept Zool.* vol. V (16), 1–192.

DAVIES, J. 1886 Parliamentary report in the *Hobart Mercury*, 5 November.

DAVIES, J.L. 1965 *Atlas of Tasmania*. Hobart: Govt Printer.

DOHERTY, K. 1977 When we caught a tiger. In *North-west Tasmania Short Stories and Articles*. Boat Harbour: N.W. Branch of the Fellowship of Australian Writers.

DUNNET, G.M. & 1974 A monograph of Australian Fleas MARDON, D.K. (Siphonaptera). *Aust. J. Zool. Supp. Ser.* vol. 30, 1–273.

— — EASTWOOD, D. 1981 Letter to the author.

EVANS, G.W. 1822 *Description of Van Diemen's Land.* London: Souter.

EWENCE, G. 1961 Article in *The Bulletin*, 9 September, p. 29.

FLEAY, D. 1946 On the trail of the marsupial wolf. *Vict. Nat.* vol. 63, 129–35; 154–59; 174–77.

FLEMING, A.L. 1939 Reports on two expeditions in search of the thylacine. *J. Soc. Preserv. Fauna Emp.* vol. 30, 20–5.

FLYNN, T.T. 1914 The Mammalian fauna of Tasmania. *Brit. Assoc. Adv. Sci. Handb.* 48–53.

FLOWER, S.S. 1931 Contribution to our knowledge of the duration of life of vertebrate animals. *Proc. zool. Soc. Lond.* 145–234.

FLOWER, W.H. 1865 On the commissures of the cerebral hemispheres of the Marsupialia and Monotremata as compared with those of Placental Mammals. *Phil. Trans. Roy. Soc. Lond.* vol. 55, 633–51.

FLOWER, W.H. 1867 On the development and succession of the teeth in the Marsupialia. *Phil. Trans. Roy. Soc. Lond.* vol. 155, 631–41.

FOLTELNY, J.G.   1967   Gibt es einen australischen Tiger? *Kosmos* vol. 63, 292–4.

GEOFFROY, St. H.E.   1810   Déscription de deux espèces du *Dasyurus. Ann. du Mus.* vol. 15, 301–6.

GILL, E.D.   1953   Distribution of the Tasmanian devil, the Tasmanian wolf and the dingo in South East Australia in Quaternary time. *Vict. Nat.* vol. 70, 86–90.

GILL, E.D.   1964   The age and origin of the Gisborne Cave. *Proc. Roy. Soc. Vict.* vol. 77, 532–3.

GLAUERT, L.   1914   The Mammoth Cave. *Rec. Western Aust. Mus.* vol. 1, 244–51.

GOLDIE, A.   1829   Report from Goldie to Curr, 13 March 1829. *Van Diemen's Land Co. Papers*, State Archives of Tasmania.

GOODRICK, J.   1977   *Life in Old Van Diemen's Land.* Adelaide: Rigby, 220 pp.

GOULD, J.   1863   *Mammals of Australia* vol. 1. London: Taylor & Francis.

GREEN, R.H.   1967   Notes on the devil (*Sarcophilus harrisii*) and the quoll (*Dasyurus viverrinus*) in north-eastern Tasmania. *Rec. Queen Vict. Mus.* vol. 27, 1–12.

GREEN, R.H.   1974   Mammals. In *Biogeography and Ecology in Tasmania*, pp. 376–96. Den Hague: Junk.

GRIFFITHS, J.,   1972   The report of the search for the thylacine. Duplicated, 19 pp.
MALLEY, J.F. &
BROWN, R.J.

GRZIMEK, B.   1976   Thylacinidae. In *Grzimek's Animal Encyclopaedia* vol. 10, 83–88.

GUILER, E.R.   1958   Observations on a population of small marsupials in Tasmania. *J. Mammal.* vol. 39, 44–58.

GUILER, E.R.   1961a   The former distribution and decline of the thylacine. *Aust. J. Sci.* vol. 23, 207–10.

GUILER, E.R.   1961b   The breeding season of the thylacine. *J. Mammal.* vol. 42, 396–7.

GUILER, E.R.   1966   In pursuit of the thylacine. *Oryx* vol. 8, 307–10.

GUILER, E.R.   1967   The fauna of Tasmania. *Tasm. Year Bk* vol. 1, 58–64.

GUILER, E.R.   1970   Observations on the Tasmanian devil, *Sarcophilus harrisii* (Dasyuridae: Marsupialia). II. Reproduction, and growth of pouch young. *Aust. J. Zool.* vol. 18, 63–70.

GUILER, E.R. 1971 The husbandry of the Potoroo. *Int. Zoo Yearbk* vol. 11, 21–2.

GUILER, E.R. 1978 Observations on the Tasmanian devil, *Sarcophilus harrisii* (Dasyuridae; Marsupialia) at Granville Harbour, 1966–75. *Pap. Roy. Soc. Tasm.* vol. 112, 161–88.

GUILER, E.R. 1982 Spatial and temporal distribution of the Tasmanian devil. *Pap. Proc. Roy. Soc. Tasm.* vol. 116, 153–64.

GUILER, E.R. & HEDDLE, R.W.L. 1970 The form of the testicular rete mirabile of marsupials. *Comp. Biochem. Physiol.* vol. 35, 415–25.

GUILER, E.R. & MELDRUM, G.K. 1958 Suspected sheep killing by the thylacine, *Thylacinus cynocephalus* (Harris). *Aust. J. Sci.* vol. 20, 214.

GUNN, R.C. 1850 Letter to the Zoological Society. *Proc. zool. Soc. Lond.* 1850, 90–1.

GUNN, R.C. 1852 A list of the mammals indigenous to Tasmania. *Pap. Proc. Roy. Soc. Tasm.* vol. 2, 77–90.

GUNN, R.C. 1863 Extracts from a letter to the Secretary of the Zoological Society. *Proc. zool. Soc. Lond.* 1863, 103–4.

HAITCH, R. 1981 Rare tiger quest. *New York Sunday Times* 29 March, p. 41.

HARRIS, G.P. 1808 Descriptions of two new species of *Didelphis* from Van Diemen's Land. *Trans. Linn. Soc. Lond.* vol. IX, 174.

HAYES, J. 1972 Account of thylacine as related by G. Stevenson. *Examiner Express.* 10 June.

HENDERSEN, J. 1832 Observations on the Colonies of New South Wales and Van Diemen's Land. Calcutta: Baptist Mission Press.

HICKMAN, V.V. 1955 The Tasmanian Tiger. *Etruscan* vol. 5 (2), 8–11.

HILL, J.P. 1900 On the foetal membranes, placentation and parturition of the native cat (*Dasyurus viverrinus*). *Anat. Anz.* vol. 18, 364–73.

HCC (Hobart City Council). 1922 Minutes—reserves committee, 18 July.

HCC. 1923 Minutes—reserves committee, 19 June, 3 July.

HCC. 1924 Minutes—reserves committee, 19 February.

HCC. 1925 Minutes—reserves committee, 21 July.

HCC. 1930 Minutes—reserves committee, 14 April.

HCC. 1935 Minutes—reserves committee, 13 February, 3 July.

HCC.	1936	Minutes—reserves committee, 16 September.
HCC.	1937	Minutes—reserves committee, 17 February, 3 March, 4 April.
Hobart Mercury.	1864	26 November, p. 4.
Hobart Mercury.	1874	Advertisement by Jemrack, 16 July.
Hobart Mercury.	1979	28 December, p. 2.
Hobart Mercury.	1980	Article, 20 March.
Hobart Mercury.	1983	News item, 23 September, p. 1.
Hobart Mercury.	1984	21 January, p. 1.
Hobart Town Gazette.	1823	Article on a thylacine incident. 2 August, p. 2.
HOCKING, G.H. & GUILER, E.R.	1983	The mammals of the Lower Gordon River region, South-West Tasmania. *Aust. Wildl. Res.* vol. 10, 1–23.
Hong Kong Standard.	1980	Tasmania Tiger eludes search. 21 September.
HOWLETT, R.M.	1960	A further discovery of *Thylacinus* at Augusta, Western Australia. *Western Australian Nat.* vol. 7, 136.
HULL, H.M.	1871	*Hints to Emigrants intending to proceed to Tasmania*. Hobart Town: Fletcher, 48 pp.
INGRAM, B.S.	1969	Sporomorphs from the desiccated carcases of mammals from Thylacine Hole, Western Australia. *Helictite* vol. 7, 62–6.
IREDALE, T. & TROUGHTON, E. le G.	1934	A checklist of the mammals recorded from Australia. *Aust. Mus. Mem.* vol. VI, 15.
JEFFREYS, C.H.	1820	Geographical and descriptive Delineations of the Island of Van Diemen's Land. London: Richardson.
JONES, R.	1970	Tasmanian Aborigines and dogs. *Mankind.* vol. 7, 256–71.
KEAST, A.	1982	The thylacine (Thylacinidae: Marsupialia) how good a pursuit carnivore? In M. Archer (ed.) *Carnivorous Marsupials*, pp. 675–84. Sydney: Roy. Zool. Soc. NSW.
KENDRICK, G.W. & PORTER, J.K.	1973	Remains of a thylacine (Marsupialia: Dasyuroidea) and other fauna from caves in the Cape Range, Western Australia. *J. Roy. Soc. Western Aust.* vol. 56, 116–22.
KIRSCH, J.A.W. & ARCHER, M.	1982	Polythetic cladistics, or when parsimony's not enough: the relationships of carnivorous marsupials. In M. Archer

(ed.) *Carnivorous Marsupials*, pp. 595–619. Sydney: Royal Zool. Soc. NSW.

KNOPWOOD, R. 1805 *The Diary of the Rev. Robert Knopwood,*
(1977) *1803–38.* Hobart: Tasmanian Historical Research Association.

KREFFT, G. 1868 Description of a new species of thylacine (*Thylacinus breviceps*). *Ann. Mag. Nat. Hist.* vol. 4 (ii), 296–7.

LAIRD, N. 1968 Article in *Hobart Mercury*, 7 October.

LARKIN, R. 1978 Article in *Hobart Mercury*, 24 February.

Launceston Examiner. 1968 News item, 26 April, p. 1.

LE FEVRE, J. 1953 Letter to the author.

LE FEVRE, P. 1953 Letter to the author.

LE SOUEF, A.S. 1926 Notes on the habits of certain families of the Order Marsupialia. *Proc. zool. Soc. Lond.* 1926, 935–7.

LE SOUEF, A.S. & 1927 *Wild Animals of Australasia.* London:
BURREL, H. Harrap.

LESTER, C. 1983 Disease takes tiger toll. *Tasm. Mail.* 30 August, p. 2.

LIGHTON, R.F. 1968 Tasmanian tiger believed sighted. *Sci. News* vol. 93, 569.

LLOYD, G.T. 1862 *Thirty-three Years in Tasmania and Victoria.* London: Houlston & Wright.

LORD, C.L. 1928 Existing Tasmanian marsupials. *Pap. Proc. Roy. Soc. Tasm.* 1927 (1928), 17–24.

LORD, C.L. & 1924 *A Synopsis of the Vertebrate Animals of*
SCOTT, H.H. *Tasmania.* Hobart: Oldham, Beddome & Meredith.

LOWRY, D.C. & 1967 Discovery of a thylacine (*Thylacinus*
LOWRY, J.W.J. *cynocephalus*) in a cave near Eucla, Western Australia. *Helictite* vol. 5 (2), 25–9.

LOWRY, J.W.J. 1972 The taxonomic status of small fossil thylacines, (Marsupialia; Thylacinidae), from Western Australia. *J. Roy. Soc. Western Aust.* vol. 55, 19–29.

LOWRY, J.W.J. & 1969 Age of the desiccated carcase of a thyla-
MERRILEES, D. cine (Marsupialia; Dasyuroidea) from Thylacine Hole, Nullarbor Region, Western Australia. *Helictite* vol. 7 (1), 15–6.

LUCAS, A.H. & 1909 *Animals of Australia.* Melbourne: Whit-
LE SOUEF, A.S. combe & Tombs.

LUKAS, J.A. 1963 Plan to hold that elusive tiger. *Toronto Globe and Mail* 24 December.

LYCETT, J. 1824 *Views in Australia.* London: J. Souter.

LYNE, A.G. 1959 The systematic and adaptive significance

of the vibrissae in the Marsupialia. *Proc. zool. Soc. Lond.* vol. 133, 79–133.

LYNE, A.G. & McMAHON, T.S. 1951 Observations of the surface structure of the hairs of Tasmanian Monotremes and Marsupials. *Pap. Proc. Roy. Soc. Tasm.* 1950 (1951), 71–84.

LYNE, J. 1886 Parliamentary report in the *Hobart Mercury*, 5 November.

LYNE, J. 1887 Parliamentary report in the *Hobart Mercury*, 27 August.

LYON, B. 1972 Letter to the author.

MACKIESON, D. 1981 Letter to the author.

MACINTOSH, N.W.G. 1975 The origin of the dingo: an enigma. In M.W. Fox (ed.) *The Wild Canids*, pp. 85–106. New York: Van Nostrand.

MALLEY, J.F. 1973 The thylacine. *Proc. Seminar—Arthur River to Pieman River.* Smithton.

MARTIN, R.M. 1836 Van Diemen's Land. In *History of Australasia*, pp. 282–85. London: J Mortimer.

MATTHEWS, L.H. 1958 Letter to the author.

MATTINGLEY, E.H. 1946 *Thylacine* and *Thylacoleo*. *Vict. Nat.* vol. 63, 143.

McADAM, E. 1973 Letter to the author.

McCANCE, J. 1970 Article in *Weekly Times*.

McNAMARA, M. 1983 Article in *Hobart Mercury*, 6 December, p. 1.

MEAD, I. 1961 Letter to the author regarding the identity of 'J.S.'

MELVILLE, P. 1833 *The Van Diemen's Land Almanak*. Hobart: J. Ross, 32–3.

MEREDITH, L.A. 1881 *Tasmanian Friends and Foes*. London: Marcus Ward.

MERRILEES, D. 1968 Man the destroyer: Late Quaternary changes in the Australian marsupial fauna. *J. Roy. Soc. Western Aust.* vol. 51, 1–24.

MERRILEES, D. 1970 A check on the radio-carbon dating of dessicated thylacine (marsupial 'wolf') and dingo tissue from Thylacine Hole, Nullarbor region, Western Australia. *Helictite* vol. 8 (2), 39–42.

MESTON, A.L. 1958 The Van Diemen's Land Company, 1825–42. *Rec. Queen Vict. Mus.* vol. 9, 1–62.

MILLIGAN, J. 1853 Remarks on the habits of the wombat, hyaena and certain species of reptiles. *Pap. Proc. Roy. Soc. Tasm.* vol. 2, 310.

MILLIGAN, J.   1859   Vocabulary of dialects of the Aboriginal tribes of Tasmania. *Pap. Proc. Roy. Soc. Tasm.* vol. 3, 239–74.

MITCHELL, P.C.   1916   Further observations on the intestinal tract of mammals. *Proc. zool. Soc. Lond.* 183–252.

MOELLER, H.   1968   Zur Frage der Parallelerscheinungen bei Metatheria und Eutheria. Vergleichende Untersuchungen an Beutelwolf und Wolf. *Z. wiss. Zool.* vol. 177 (3/4), 283–392.

MOELLER, H.   1970   Vergleichende Untersuchungen zum Evolutionsgrad der Gehirne grosser Rabbeutler (*Thylacinus, Sarcophilus* und *Dasyurus*). *Z. f.zoologische Syst. und Evol. Fors.* vol. 8, 69–88.

MOLLISON, B.C.   1951   Statements attributed to his grandfather. Queen Vict. Mus. files, quoted in S.J. Smith 1980.

MORRIS, D.   1962   The thylacine re-discovered. *Proc. zool. Soc. Lond.* vol. 138, 4, 668.

MUDIE, R.   1829   The Picture of Australia: exhibiting New Holland, Van Diemen's Land and all the settlements from the first at Sydney to the last at Swan River. London: Whittaker, Treacher.

MULVANEY, J.D.   1969   *The prehistory of Australia.* London: Thames & Hudson.

MUNDAY, B.L. & GREEN, R.H.   1972   Parasites of Tasmanian Fauna. II. Helminthes. *Rec. Queen Vict. Mus.* vol 44, 1–15.

NICHOLLS, W.   1960   Letter to the author.

OWEN, R.   1837   On the structure of the brain in marsupial animals. *Phil. Trans. Roy. Soc. Lond.* 87–96.

OWEN, R.   1841   *Marsupialia. Todd's Cyclopaedia of Anatomy* vol. 3, 257–81.

OWEN, R.   1843   On the rudimentary marsupial bones in the *Thylacinus. Proc. zool. Soc. Lond.* vol. 11, 148–9.

OWEN, R.   1846   On the rudimental marsupial bones in the *Thylacinus. Tasm. J.* vol. 2, 447–9.

OWEN, R.   1868   *Anatomy of Vertebrates* vol. III. London: Longmans Green.

OWEN, R.   1877   *Researches on the fossil remains of Australia, with a notice of the extinct marsupials of England.* London: the author.

PACKER, H. 1970 Is it the Tasmanian tiger? *Australian Women's Weekly* February 25.

PADMAN, W. 1949 Letter to the author.

PARKER, H.W. 1833 *The rise, progress and present status of Van Diemen's Land.* London: J. Cross.

PARTRIDGE, J. 1967 A 3,300 year old thylacine (Marsupialia: Thylacinidae) from the Nullarbor Plain, Western Australia. *J. Roy. Soc. Western Aust.* vol. 50, 57–9.

PATERSON, W. 1805 Report in *Sydney Gazette and New South Wales Advertiser* vol. 3, 24 April.

PEARSE, A.M. 1981 Aspects of the biology of *Uropyslla tasmanica.* MSc thesis, University of Tasmania.

PEARSON, J. & de BAVAY, J.M.⁻ 1953 The urogenital system of the Dasyurinae and Thylacininae (Marsupialia: Dasyuridae). *Pap. Proc. Roy. Soc. Tasm.* vol. 87, 175–99.

PLOMLEY, N.J.B. 1966 *Friendly Mission.* The Journals of G.A. Robinson. Hobart: Tasmanian Historical Research Association.

POCOCK, R.I. 1914 On the facial vibrissae of mammalia. *Proc. zool. Soc. Lond.* 1914, 889–912.

POCOCK, R.I. 1926 The external characters of *Thylacinus, Sarcophilus* and some related marsupials. *Proc. zool. Soc. Lond.* 1926, 1037–84.

Pretoria News. 1981 Search on for tiger with a pouch. 24 October, p. 8.

RANSON, B.H. 1905 Tapeworm cysts (*Dithyridium cynocephali* n. sp.) in the muscles of a marsupial wolf (*Thylacinus cynocephalus*). *Trans. Amer. Micros. Soc.* vol. 27, 31–2.

RENSHAW, G. 1938 The thylacine. *J. Soc. Preserv. Fauna Emp.* vol. 35, 47–9.

RIDE, W.D.L. 1964 A review of Australian fossil marsupials. *J. Roy. Soc. West. Aust.* vol. 47, 97–131.

RIDE, W.D.L. 1970 *A guide to the native animals of Australia.* Melbourne: Oxford University Press.

RITCHIE, B. 1961 Letter to the author.

ROBERTS, M.G. 1915 The keeping and breeding of Tasmanian devils (*Sarcophilus harrisii*). *Proc. zool. Soc. Lond.* 1915, 575–81.

ROSS, J. 1830 *The Hobart Town Almanack.* Hobart: J. Ross.

ROTH, H.L. 1899 *The Aborigines of Tasmania.* Hobart: Fuller.

ROUNSEVELL, D.E. 1983 Thylacine. In *The Complete Book of Australian Mammals.* Sydney: Australian Museum.

ROUNSEVELL, D.E. & SMITH, S.J. — 1980 — Recent alleged sightings of the thylacine in Tasmania. Paper read at the Aust. Mammal Soc. Conf.

J.S. — 1862 — Tasmanian tigers. Letter to *Launceston Examiner*, 22 November.

SACK, M. — 1981 — Letter to the author.

SARICH, V., LOWENSTEIN, J.M. & RICHARDSON, B.J. — 1982 — Phylogenetic relationships of *Thylacinus* (Marsupialia) as reflected in comparative serology. In M. Archer (ed.) *Carnivorous Marsupials* vol. 2, pp. 707–19. Sydney: Roy. Zool. Soc. NSW.

Saturday Evening Mercury (Hobart). — 1984 — 21 January, p. 1.

SAUNDERS, J. — 1972 — Letter to the author.

SAWLEY, F. — 1980 — Letter to the author.

SAYLES, J. — 1980 — Stalking the Tasmanian tiger. *Anim. Kingd.* vol. 82 (6), 35–40.

SCOTT, P. — 1965a — Land Settlement. *Atlas of Tasmania*, pp. 43–5. Hobart: Govt Printer.

SCOTT, P. — 1965b — Farming. *Atlas of Tasmania*, pp. 58–65. Hobart: Govt Printer.

SHARLAND, M.S.R. — 1939 — In search of the thylacine. *Proc. Roy. Soc. NSW* 1938/39 (1939), 20–36.

SHARLAND, M.S.R. — 1957 — In search of the vanished 'tiger'. *People* 3 April, 25–6.

SHARLAND, M.S.R. — 1966 — *Tasmania.* Sydney: Nelson, Doubleday.

SILVER, S.W. — 1874 — *Handbook for Australia and New Zealand.* London: Silver & Co.

SIMPSON, G.G. — 1941 — The affinities of the Borhyaenidae. *Amer. Mus. Novit.* vol. 1118, 1–6.

SKEMP, J.R. — 1958 — *Tasmania Yesterday and Today.* Melbourne: Macmillan.

SMITH, G. — 1909 — *A naturalist in Tasmania.* Oxford: Clarendon Press.

SMITH, S.J. — 1980 — The Tasmanian Tiger—1980. Hobart: Tas. National Parks Wildl. Serv.

SPICER, J. — 1978 — Letter to the author.

SPRENT, J.F.A. — 1971 — A new genus and species of Ascaridoid nematode from the marsupial wolf (*Thylacinus cynocephalus*). *Parasitol.* vol. 63, 37–43.

SPRENT, J.F.A. — 1972 — *Cotylascaris thylacini*, a synonym of *Ascaridia columbae*. *Parasitol.* vol. 64, 331–2.

STEPHENSON, N.G. — 1963 — Growth gradients among fossil monotremes and marsupials. *Palaeontol.* vol. 6, 615–24.

STEVENSON, G. — 1972 — *Sunday Examiner-Express*, 10 June.

SUMMERS. R.G.   1937   Animals and Birds Protection Board file
                                H/60/34.

SWAINSON, W.   1864   In H. Murray, *Encyclopedia of Geography*.
                                London: Longmans, Green.

TATE, G.H.H.   1947   On the anatomy and classification of the
                                Dasyuridae (Marsupialia). *Bull. Amer.
                                Mus. Nat. Hist.* vol. 88, 97–156.

TEMMINCK, G.R.   1824   *Monogr. Mamm.* 1 (XXIII), 23; 60; 267.

TERRY, E.V.   1961   Letter to the author.

THOMAS, O.   1888   *Catalogue of the Marsupialia and Monotre-
                                mata in the collection of the British Museum
                                (Natural   History)*.   London: British
                                Museum.

The Times (London).   1979   Naturalists hunt extinct tiger in Tasma-
                                nia. 16 November.

TROUGHTON, E.   1941   *Furred Animals of Australia.* Sydney:
le G.   Angas & Robertson.

TYNDALE-BISCOE,   1973   *Life of Marsupials.* London: Arnold.
C.H.

van DEUSEN, H.M.   1963   First New Guinea record of *Thylacinus*.
                                J. Mammal. 44, 279–80.

VECHTMANN, N.   1980   Hoe 'uitgestorven' is Tasmaanse buidel-
                                wolf? *Het Vrije Volk* 13 June.

VRYDAGH, J.M.,   1964   *Des fossiles de demain. Trieze mammifères
CARAM, M. &   menacés d'extinction.*   Brussels: Pro
PETTER, F.   Natura Union Int. Prot. Nature.

WALSH, H.   1979   Letter to the author.

WARLOW, W.   1833   Systematically arranged catalogue of the
                                mammals and birds belonging to the
                                Museum of the Asiatic Society, Calcut-
                                ta. *J. Asiatic Soc. Bengal* vol. 2, 97–100.

Washington Post.   1980   Tasmanian tiger sought in Australia. 29
                                May.

Weekly Times.   1971   8 September, p. 38.

WEDGE, J.H.   1962   *The Diaries of John Hilder Wedge, 1824–
                                35.* G. Crawford *et al.* (eds). Hobart:
                                Royal Soc. of Tas.

WEST, J.   1852   *History   of   Tasmania.*   Launceston:
                                Dowling.

WHITLEY, G.P.   1973   I remember the thylacine. *Koolewong*
                                vol. 2 (4), 10–1.

WIBER, J.   1977   Letter to the author.

WIDOWSON, H.   1829   *The Present State of Van Diemen's Land.*
                                London: Robinson.

WILFORD, J.N.   1980   A new search for rare tiger. *Intern.
                                Herald Tribune* 6 June.

WILLETT, J.   1927   Article in the *Hobart Mercury* 13
                                September.

WOODBURNE, M.O.   1967   The Alcoota Fauna, central Australia: an integrated palaeontological and geological study. *Aust. Bur. Min. Resources, Geol. Geophys. Bull.* 87.

WOOD JONES, F.   1921   The status of the dingo. *Trans. Roy. Soc. S. Aust.* vol. 45, 254–63.

WOOD JONES, F.   1929   *Man's place among the mammals.* London: Arnold.

WOODS, S.   1980   Letter to the author.

WRIGHT, E.P.   1892   Family LXXII—The Dasyures. In *Concise Natural History.* London: Cassells.

ZUCKERMAN, S.   1953   The breeding season of mammals in captivity. *Proc. zool. Soc. Lond.* vol. 122, 827–950.

Note

Index